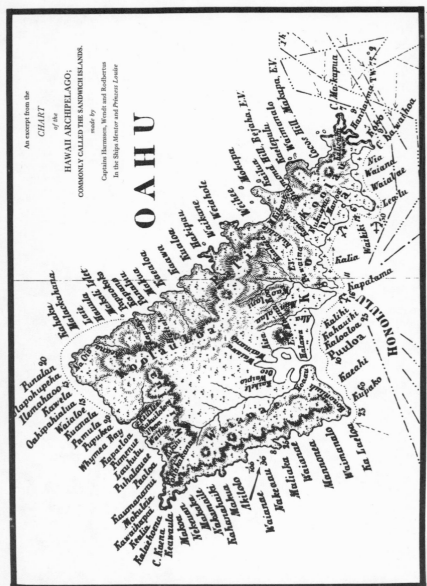

(courtesy of the Bishop Museum Library)

A BOTANIST'S VISIT TO OAHU IN 1831

Being the Journal of Dr. F. J. F. Meyen's travels
and Observations about the Island of Oahu

The text is an excerpt from the original publication:
Reise um die Erde Aüsgefuhrt auf dem Königlich
Preussischen Seehandlungs-Schiffe, Prinzess Louise,
Commandirt von Capitain W. Wendt
In den Jahren 1830, 1831 und 1832

Translated by Astrid Jackson

Edited by Mary Anne Pultz

PRESS PACIFICA, LTD.
1981

Meyen, F.J.F. (Franz Julius Ferdinand), 1804-1840.
    A botanist's visit to Hawaii in 1831.

Excerpt from the orginal publication: Reise um die Erde
ausgeführt auf dem Königlich Preussischen Seehandlungs-
Schiffe Prinzess Louise, commandirt von Capitain W. Wendt,
in den Jahren 1830, 1831, und 1832.
    Includes bibliographical references and index.
    1. Botany—Hawaii—Oahu.   2. Meyen, F.J.F. (Franz
Julius Ferdinand), 1804-1840. 3. Oahu (Hawaii)—Description
and travel. 4 Botanists—Germany—Biography. I. Pultz, Mary
Anne. II. Title.
QK473.H4M49213   581.9969'3   81-7353
ISBN 0-916630-23-4   AACR2

Cover drawing by Katherine Livermore courtesy of the Bishop
Museum.

Maunufactured in the United States of America.

Published by:
    Press Pacifica Ltd., P.O. Box 1227, Kailua, Hawaii, 96734.

TABLE OF CONTENTS

EDITOR'S NOTE

The German text written by a Prussian botanist visiting Oahu 150 years ago inevitably contains a number of outdated but interesting renderings, both of scientific and of Hawaiian names. The updating of scientific plant names is dealt with using Dr. Harold St. John's series of botanical notes, which have been numbered and appear in the Appendix. The only editorial changes of scientific names in the text have been the occasional correction of typographical errors in the original Latin names, and the uniform decapitalization of all capitalized species epithets so that those scientific names which are still in use will appear in the text in a standard modern form.

Plant names are not the only terms that have been changed and standardized since Meyen's time. Meyen seems to have noted down many Hawaiian words as his German ear heard them, such as his rendering of Aloha as "Arocha." For the sake of historical interest, Meyen's spellings have been used at the first appearance of each word, with the modern spelling appended in brackets; but on subsequent appearances, the standard modern Hawaiian spellings have been substituted throughout the text.

Another curiosity is Meyen's use of the word "Indians" to denote native Hawaiians. Although the use of this word has an odd ring today, in Meyen's time it probably seemed quite natural to use the same term for the natives of Hawaii as for the natives of South America, which he had visited on an earlier phase of his voyage.

Three sets of notes appear throughout the text:

* —Meyen's original footnotes
† —editorial footnotes
1 —Harold St. John's botanical notes

ACKNOWLEDGMENTS

The Pacific Translators is a committee dedicated to making foreign-language sources of literature about the Pacific area available in translation to the English-reading public. Until the publication of this volume, F. J. F. Meyen's account of his early botanizing visit to Oahu was available in Hawaii only as a section of a rare three-volume German work in the Bishop Museum library detailing Meyen's voyage around the globe. Encouraged by former Bishop Museum librarian Margaret Titcomb, the Pacific Translators took the initiative, first, to have the work translated by Astrid Jackson, and then, to see it through the many steps required to produce a valuable and enjoyable English edition. Special assistance was provided by Cynthia Timberlake in the use of Bishop Museum library resources, and by Margaret Titcomb in compiling the index.

To make Meyen's nineteenth-century botanizing intelligible to those who are familiar with the plants by their Hawaiian and modern scientific names, Dr. Harold St. John has generously assisted by annotating the scientific and Hawaiian names of the many plants which have been renamed since Meyen's time. Of additional interest, Dr. St. John points out several cases in which Meyen must have been mistaken in his plant identification. Along similar lines, H. Douglas Pratt has provided information concerning the probable identity of the bird that Meyen sighted in Nuuanu valley.

The book is introduced by Dr. E. Alison Kay, a University of Hawaii professor who is concerned with Hawaii's natural history both as a scientist and as an educator. Dr. Kay has kindly provided a historical framework for appreciating Meyen's account, as well as discussing some of the highlights of his scientific work which are of enduring interest today.

INTRODUCTION

E. Alison Kay

When the *Prinzess Louise*, of Prussian registry and under the command of Captain Wendt, approached the south shore of Oahu on June 24, 1831, she dropped anchor not off the reef at Waikiki as her predecessors had done thirty years earlier, but in the calm waters inside the reef at Honolulu Harbor. The ship was on the third Prussian voyage around the world, her mission to emulate for Prussia the deeds and exploits of the explorers and naturalists of Britain, France, and Russia who for more than half a century had sailed the Pacific. Aboard the *Prinzess Louise* as a member of the ship's company was Franz Julius Ferdinand Meyen, 27 years old, with a medical degree from the University of Berlin, and charged, on recommendation of Alexander von Humboldt, with the natural history researches of the voyage. Meyen's account of Oahu, seen during the visit between June 24 and June 30, 1831, is a perceptive description of natural history, culture and society.

Oahu was first visited by foreigners in March 1779 when Captains King and Clerke of the *Resolution* and *Discovery*, following Captain Cook's death at Kealakekua Bay, stopped briefly for fresh water off Waimea Bay, where a small party went ashore. Seven years later, in June 1786 (and again in November 1786 and September 1787) Oahu was again visited by foreigners when Captains Portlock and Dixon of the *King George* and *Queen Charlotte* sailed along the southern coast and put in for water at Waialae (King George's Bay) and off Waikiki (Queen Charlotte's Bay). Dixon briefly described Waikiki, ". . . crouded with new plantations laid out in a most regular order, and which seem to be in a very flourishing state of cultivation."[1] In 1792 Captain Vancouver and the naturalist of his ship, James Menzies, also saw Oahu from outside the reef at Waikiki. Vancouver wrote that "On the shores [of the bay] the villages appeared numerous and in good repair, and the surrounding country interspersed with deep, though not extensive valleys, which with the plains near the seaside presented a high degree of cultivation and fertility . . . to the northward through the village . . . an exceedingly well-made causeway . . . opened our view to a spacious plain, which the major part divided into fields of irregular shape and figure . . . planted with taro root, in different stages of inundation."[2] Menzies echoed Vancouver's description, but added that "The verge of the shore [of Waikiki] was planted with a large grove of coconut palms. . ." and that ". . . the plantation . . . was laid out with great neatness into little fields planted with taro, yams, sweet potatoes and cloth plant."[3]

Sixteen years later, when Archibald Campbell reached Oahu aboard the Russian ship *Neva* in 1809, the ship was carefully maneuvered through a channel in the reef, and anchored in the "harbor of Hanaroora."[4] Honolulu had displaced Waikiki as the "center" of Oahu not only because the king had removed the royal residence from Waikiki to Honolulu, but because a safe harbor, "Fair Haven," had been discovered in 1793 when Captain Brown took his ships through a narrow channel in the reef. Campbell described Honolulu as consisting of several hundred houses, "well shaded with large cocoa-nut trees. . ." and he also described Wymumme [Pearl Harbor], "an inlet of the sea about twelve miles to the west of Hanaroora. . . ."[4] On his way to Wymumme, Campbell ". . . passed by footpaths, winding through an extensive and fertile plain, the whole of which is in the highest state of cultivation. Every stream was carefully embanked to supply water for the taro beds. Where there was no water, the land was under crops of yams and sweet potatoes. The roads and numerous houses are shaded by cocoa-nut trees, and the sides of the mountains covered with wood to a great height."[4]

By 1820, Oahu's landmarks were well known. Hiram Bingham spent part of his first day in Honolulu, April 14, 1820, sight-seeing, and, having climbed to the rim of Punchbowl, described "a beautiful view":

"On the east were the plain and groves of Waikiki, with its ampitheatre of hills, the southeastern of which is Diamond Hill. . . . Below us, on the south and west, spread the plain of Honolulu, having its fish-ponds and salt making pools along the seashore, the village and fort between us and the harbor, and valley stretch-ing a few miles north into the interior, [with] habita-tions and numerous beds of *kalo*. . . . Through this valley, several streams descending from the mountains in the interior, wind their way. . . . From Diamond Hill on the east, to Barber's Point, and the mountains of Waianae, on the west, lay the seaboard plain, some twenty-five miles in length, which embraces the vol-canic hills of Moanalua, two or three hundred feet high, and among them, a singular little lake of sea-water. . . the ravine of Moanalua, the lagoon of Ewa, and nu-merous little plantations and hamlets, scattered trees and cocoanut groves. A range of mountains, three or four thousand feet high, stretches from the northwest-ern to the eastern extremity of the island. Konakuanui, the highest peak, rises back of Punchbowl Hill, and north by east from Honolulu, eight miles distant, and four thousand feet high, often touching or sustaining, as it were, a cloud."[5]

The itinerary followed by Meyen in 1831 was almost precisely the scene described by Bingham. The young naturalist during the course of five days travelled on foot and on horseback up Nuuanu Valley to the Pali (and down to the windward side not

described by Bingham), up to Punchbowl and Sugarloaf, across
the taro patches of Waikiki to Diamond Head and into the crater,
and finally out across the plain of Honolulu through the royal
fishponds to Pearl Harbor, enroute to which he saw Salt Lake,
and, in the distance "the mountains of Waianae." Meyen must
have covered at least 185 miles on the five days of his explorations.

Meyen was perhaps more perceptive than the usual traveller,
for, as a physician,he was a trained and thoughtful observer, and
his special training in "pharmacie" endowed him with a strong
botanical background. Indeed, he was author of two volumes of
plant anatomy before he sailed on the *Prinzess Louise: Sur les
matieres contenues dans les cellules des vegetaux* (Berlin, 1828)
and *Sur la phytotomie* (Berlin, 1830). His descriptions of Oahu
suggest he was also something of a geologist, an anthropologist,
and a social commentator.

Meyen's description of the vegetation is of surpassing interest.
Other botanists had visited Hawaii. In 1815, when the Russian
ship *Rurick*, commanded by Captain Kotzebue was in the islands,
Adelbert Chamisso and Frederick Escholtz botanized on Oahu
between November 27 and December 8 (and again on a second
visit in 1817). D. Gaudichaud, aboard the French ship *Uranie*,
wrote of "la vegetation extraordinaire." Meyen's descriptions are
especially discerning, however, for he noted not only differences
in composition but the heights at which changes in vegetation
occurred. On the plain of Honolulu he described taro and other
agricultural crops, and then he noted that taro and bananas
disappeared at about 800 feet when he entered the rain forest
of upper Nuuanu Valley. At 1200 feet he found another change,
"the physiognomy of the vegetation changed from *Pandanus*
to little bunches of *Peperomia, Plantago* and *Oxalis.*" Again on
his excursion to Puowaina [Punchbowl] and Sugarloaf,he noted
the heights at which the vegetation changed: the flat valley
through which he hiked was completely barren up to an elevation
of 600 to 700 feet, and then he walked through a meadow of
Sida, through a fern forest, and finally among lobeliads, "Nowhere on the island did we find more of those strange Lobeliaceae which Mr. Gaudichaud described, than here."

Aspects of Hawaii's unique biota had already been recognized when Meyen visited Oahu. Animals and plants, birds, tree snails, lobeliads, and ferns had been described and specimens preserved in museums and herbaria in Europe. Meyen saw another unique feature of the biota. Comparing it with that he had seen in Brazil, he gave pride of place to the Hawaiian land snails, not as objects of taxonomic interest but for their role in the forests, "The forests of Brazil abound with ugly amphibia and countless insects . . . . Here on the Sandwich Islands nature has placed countless land snails instead of insects on the leaves of trees."

Other aspects of Oahu's natural history were noted and commented on by Meyen. Arriving offshore of Honolulu, he recorded the heights of the tides and remarked on the coral reef which appeared to surround Oahu. Not only did he detail the depth and width of the reef, but he also speculated on the origins of coral islands. Later, as he journeyed through valleys and climbed the mountains of Oahu, he described rocks and other geological features.

Meyen's detailed descriptions of planting practices (including descriptions of the dimensions of taro patches) and the uses of various plants, from kukui (for lamps) to rushes (for fine mats) and the legume *Tephrosia* as a narcotic for fish, and his comments on pollination and seeds are of particular interest. Ethnobotanists have long recognized that taro was not seeded and that the Hawaiians depended on cutting to propagate their crops. Meyen wrote that "Just as our cultivated plants, which are grown primarily for their roots, only seldom bear flowers and fruits, so with taro." Others of his observations are percipient: seeing a woman transferring the pollen of poppy plants to the stigma he asked if the Hawaiians had not surmised sex differentiation earlier than the Europeans. (Handy and Handy make note of Meyen's comments.[6])

Through Meyen we meet Kuakini (John Adams); Kauike-'aouli; Ka'ahumanu, and Kinau. It is disappointing that he does not say more of Dr. Rooke who accompanied him on some of his excursions, for the English doctor was himself a knowledgeable naturalist.

Meyen praises and he criticizes. He is impressed by Marin and the agricultural produce in the markets. He is critical of the missionaries in several instances, and of the practices of the royal family as they concerned the economy of the kingdom.

It is a truism that the fifty years of Hawaiian history after the visit of Captain Cook were years of turmoil and change. Just how extensive that change was in terms of landscape is not well documented. The early visitors were most concerned with the agriculture of the lowlands, as Dixon and Vancouver described the inland regions of Waikiki. Few travellers described the vegetation of even the lowland hills. Meyen provides some insight. He was ecstatic over the profuse vegetation of the rain forest, "Nowhere . . . did we see such a charming picture of nature . . . the greatest profusion of the gayest tropical vegetation. . . ." Yet in nearly the same breath he ". . . saw the mountains everywhere covered with grazing horses and horned cattle. The island of Oahu has more than 2000 head," and added that ". . . the western range . . . is the only place where one can still find some sandalwood."

It was those cattle as well as the inroads of urbanization that in subsequent years drastically changed the landscape so vividly described by Meyen. Not only have the taro patches gone from the "Valley of Honolulu," but so too have the masses of fern, the "splendid meadows" of *Sida* and "those strange Lobeliaceae" gone from Sugarloaf. What remain are the careful and percipient comments of a young man who not only described but appreciated such beauty, as he himself wrote five years after he had returned to Europe, "Nature in each zone of the earth has peculiar beauties, whether it be in the sunny isles of the South Sea, under the cool shade of the northern oaks in the lovely well-watered mountain valleys, on the picturesque glaciers of the higher ranges, or in the midst of the Lybian deserts. But nature is silent to the unobservant man, and that rich spring of enjoyment escapes him, which has the power to delight and cheer us, even when suffering from the severest blows of fate."[7]

Returning to Europe on the *Prinzess Louise,* Meyen became professor of botany at the University of Berlin, publishing several volumes as the result of his trip around the world, among them the *Grundriss der Pflanzengeographie* in 1836, and the three-volume *Neues System der Pflanzen-Physiologie* in 1837-1839. Meyen died in Germany at the age of 36 in 1840.

References Cited

1. Dixon, George. *A voyage around the world . . . 1785-1788.* London, 1789.
2. Vancouver, George. *A voyage of discovery to the South Pacific Ocean . . . 1790-1795.* 3 vols. London, 1798.
3. Menzies, Archibald. *Hawaii Nei 128 years ago.* Honolulu, 1920.
4. Campbell, Archibald. *A voyage round the world from 1806 to 1812.* University Press of Hawaii, 1967.
5. Bingham, Hiram. *A residence of twenty-one years in the Sandwich Islands.* Hartford, 1847.
6. Handy, E. S. Craighill and Elizabeth Green Handy. *Native planters in old Hawaii.* B. P. Bishop Museum, Honolulu, Hawaii, 1972.
7. Meyen, Franz J. F. *Outline of the geography of plants.* Translated by Margaret Johnston. London, 1846.

Arrival in Honolulu

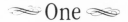 One

Our Stay on the Island of Oahu
and Comments about
the Political Situation of the Sandwich Islands

We had hardly dropped anchor in the roadstead of Honolulu —the capital city and royal residence of the Sandwich
Islands—when several merchants came on board the *Prinzess* and
greeted us as old friends, since the ship had already visited these
beautiful islands once before. Soon thereafter we received a visit
from Kuakini, present Governor of the island of Oahu, who is
known by the name of John Adams. He came in a large sailboat
and called on us out of curiosity. It was the same man with
whom we had become acquainted earlier through Mr. Hofmann's
account of Mr. von Kotzebue's second voyage.[†] At that time he
was Governor of Hawaii. The giant size and shapeless figure of
this man at first filled us with amazement. His body is so large
and so ungainly, that he cannot stand without support for one
minute but must sit down immediately or at least lean against
something. Kuakini was not able to climb on board the ship and
so had to be pulled up with a heavy rope which was wound
around his body. As soon as Kuakini had attained a firm footing

[†]Kotzebue, Otto von. *A Voyage of Discovery, . . . in the years 1815-1818, . . . in
the ship Rurick. . . .* London: Longman, Hurst, Rees, Orme, and Brown. 1821. 3 v.

on board the *Prinzess,* he looked around with the greatest indifference and said almost nothing. The extraordinarily large, heavy face with its dark red, coarse skin, thick, protruding lips, terribly broad nose and large, reddish eyes gave the man a frightful appearance.

Captain Wendt informed the Governor that His Majesty the King of Prussia had sent a great number of gifts to the ruler of the Sandwich Islands which they had on board the *Prinzess Louise* and that they had come to deliver them. Kuakini said little in reply but assumed a very pensive expression and fell into deep thought. Then he paid his respects and, after having had several glasses of wine, left the ship.

We had lain at anchor for over an hour, already the visiting merchants had left us and the Governor had returned to land but neither the canoes nor the swimming nymphs appeared which once used to surround visiting ships in great numbers. Finally a single vessel appeared with two Indians on board, but it did not approach until it had been called several times. The two Indians brought coconuts and watermelons which they spread out on board our ship and offered for sale. They were completely naked except for the marro [malo], namely that narrow piece of material which they had wound around their hips. We were even more surprised, however, when they demanded the great sum of 2 Spanish taler (3 Prussian Taler) for three watermelons and seven coconuts and would not part with them for under 9 real da plata (1 Taler and 24 Prussian Silbergroschen). So far we had not set foot on the Sandwich Islands; as yet we knew little about the life and ways of the missionaries who at this time were oppressing these blessed islands, but from this outrageous increase in the price of food we concluded that times must have changed tremendously on the Sandwich Islands. It was no longer a matter of trading for nails or for other bits of iron or of bartering for old pieces of clothing. It was money alone, and Spanish silver at that, which these poor people were after.

While the Indians were selling their fruit, the boat which they had tied to the ship broke loose. Immediately one of the

Indians jumped overboard and brought it back. He swam, imitating the motions of frogs, employing exactly the same method as is presently being used to teach swimming in northern Europe.

About one and one half hours after Kuakini had left the ship the flag of the Sandwich Islands was raised on the fort of Honolulu and now the *Prinzess Louise* gave a seventeen-gun salute while the royal standard was raised upon her mainmast. Immediately the fort of Honolulu replied, likewise with a seventeen-gun salute.

In the afternoon Captain Wendt and I went on shore in order to immediately deliver the letter from His Majesty the King of Prussia to the ruler of the Sandwich Islands.

We already knew from the chart in Mr. von Krusenstern's[†] atlas, as well as from Mr. Hofmann's[*] accounts, that the entire harbor of Honolulu is formed by coral reefs which have left only a very narrow entrance. Still, this entrance has enough water even for larger vessels, at least at high tide. It is regrettable that this map of the harbor does not also present a sketch of the roadstead since all the larger vessels must lie there initially.

On the average the tide at the Sandwich Islands attains a height of three feet only at very high tide. But often several weeks pass during which the tide is quite unnoticeable—this depending entirely on the strength and direction of the trade winds. Larger vessels can enter the harbor only at high tide. The harbor itself is excellent, however, and ships can be moored there in complete safety. At the side of the entrance into the harbor they have fastened three white flags on large floating wooden rafts so that the ships will not miss the way. If one strays even slightly from the direction of these signs, one comes immediately upon the coral reef where even the lightest boats run aground and are damaged.

The island of Oahu, as far as we were able to see, is surrounded by this great coral reef. In some places, for instance

[*] Karsten, C. J. B. *Archiv. F. Bergbau*, Vol.I, Issue 2, 1818, p. 299.

[†] Krusenstern, A. J. von, *Voyage around the world in the years 1803, 1804, 1805, & 1806, by order of His Imperial Majesty Alexander the First, on board the ships Nadeshda and Neva . . .* (2 vols, London, 1813).

right in front of Honolulu, it forms high, wide flats which are left completely dry at ebb tide and stretch far out into the sea. Here the convicts work at quarrying the stone which is used for public buildings. The stone is hewn from the reef in sections a foot and a half long, 8 to 9 inches wide and just as thick. Then it is either carried away singly on the heads of the Indians or is bound together in twos, fastened to a pole and in this way carried onto the land.[*] Around Oahu, just as around other islands of the South Seas, the coral reef forms an enclosure which drops off very suddenly at a certain distance from the coast. At this edge, even when the sea is quite calm, there is a constant, strong surf which warns the seafarer who is unfamiliar with the area of the danger—often just in time. The corals which form the reef of Oahu are mainly *madreporae* and *caryophyllae*.[†] We have never seen them alive and still covered with polyps but this is due only to the shortness of our stay on Oahu and to the numerous duties to which we were obligated there. Usually the coral reefs at the Sandwich Islands lie 2 to 3 feet under water at ebb tide. At their steep slopes, however, they drop off into the depths. We have found their continuation at 15 fathoms and as much as a half mile from the land. From this we can probably assume—even though some recent voyagers have advanced a contrary opinion —that these corals often build up from enormous depths and indeed have done so here. Of course they do not build islands, as Mr. von Chamisso[††] and others have already shown, but they do cover the underwater slopes of an island with crusts of varying thickness. Trying to determine the depth from which they originate would always yield very uncertain results. Most of the

[*] Note: In former times it was very common for sailors from ships which stopped at the Sandwich Islands to run away in order to stay behind there. At the present time no one is allowed to settle on the Sandwich Islands without the permission of the government and sailors who defy this law are sentenced to labor in the quarries on the coral reef.

[†] *Madreporae* and *caryophyllae*. Terms used in older classifications to include the stony corals.

[††] Chamisso, Adelbert von, 1781-1838, was a naturalist on the expedition with von Kotzebue, the author of *A voyage of discovery into the South Seas and Bering Straits.*

small islands which lie scattered about the South Seas are the tops of mountains, usually of volcanic origin, which rise up from the depths of the ocean. There is no doubt that through the building of the corals many of these islands will increase in size within the sea and that this increment in size will become noticeable in the coming centuries.

After a half-hour drive we arrived in the vicinity of the fort and were greeted by thousands of Indians whose "Arocha! Arocha!" [Aloha!] rang from all sides. The Governor had immediately spread the news of the gifts which we had brought for the ruler of the Sandwich Islands throughout Honolulu and the people had been assembled to receive us with much ceremony. All points of the shore, wherever we looked, were covered with Indians who, in a happy confusion, young and old, men and women, awaited us here and thronged around us. Everything that we saw here was new; everything surprised us. We were now entering upon that land which, since our first reading of Cook's *Voyages,*[†] had filled our fantasy with the liveliest images and we were seeing our expectations fulfilled. The entire assembly accompanied us with loud jubilation and a fearful pushing and shoving to the house where we were to stay. The procession grew larger and larger as it proceeded because crowds of people streamed in from all sides. Now babbling youngsters ran before us, now friendly old men extended their hands to us, and young girls and women greeted us from a distance. But what a sight this colorful crowd was! Among all these Indians there were not two who wore similar articles of clothing. Most of them were almost completely naked; a malo was bound around their hips and a tapa, i.e., a larger piece of material which reached down to the thighs, was draped over the shoulders. Most of the women

[†]Cook, James, and King, James. *A Voyage to the Pacific Ocean . . . performed under the direction of Captains Cook, Clerke, and Gore, in his Majesty's ships the Resolution and Discovery in the years 1776, 1777, 1778, 1779, and 1780.* Vols. I and II by Cook, vol. III by King. Published by order . . . of the Admiralty. First Edition. London: Printed by W. and A. Strahan for G. Nichol and T. Cadell. 1784. 3 v. and atlas.

wore a simple, loose garment which is tied at the neck and known by the name of "missionary shirt."

But the friendly reception we received here was no longer that outpouring of innocent curiosity, that free impulsiveness, which had in earlier times greeted travellers. Rather it was a gathering arranged by orders of the Governor, and the poor people had to be happy because this was an occasion on which they were permitted to be happy.

Soon after our arrival we received a visit from Governor Kuakini, during which a large crowd of curious Indians surrounded our house and even came into the room in which we lived, staring at us constantly. The Governor informed us that Kauike-aouli [Kauike'aouli], the young ruler of the Sandwich Islands, was not at home, that he had been staying at a country estate three miles from Honolulu for several days, but that he had been sent for immediately and it was hoped that he would return to the palace today.

In the afternoon we used our time to look at the city of Honolulu and had a Spanish merchant, who was a resident of the city, take us to the famous missionary Bingham, for whom we had letters from Europe. On our way to Mr. Bingham's house we encountered a very distressing spectacle which lowered our esteem of the missionaries right at the outset. What we saw was two missionary wives taking an outing, sitting in a little carriage drawn by several Indians.

Perhaps it will appear inappropriate to some of our good readers if we, in the course of this narrative, present a considerable amount of information which pertains only to the private life of the missionaries. It did seem necessary to us, however, to collect all such facts so that it might be easier to become acquainted with the activities of these men without actually having seen them. The missionaries of the South Sea Islands are no longer private persons. They have drawn the attention of the entire civilized world upon them and everyone demands an account of their dealings. The missionaries of the Sandwich Islands are North Americans and it is they alone who are being severely criticized from all sides. They have undermined the prosperity of the land instead of furthering it. They have driven out hospitality—one of the most beautiful characteristics of man in his natural state—as well as joyfulness from these happy islands and have introduced in their place a religion which the Indians cannot understand. There are men who have come forward, interestingly enough, in regions farthest away from the scene of action, men who were unfamiliar with what had so far been written about this situation, who nevertheless set out to defend the dealings of the missionaries on the Sandwich Islands with the greatest fervor. Indeed, toward this end they sometimes employed methods which seemed partly inadmissible, partly ridiculous. We would enter into greater detail if that man who most zealously defended these missionaries were still among us but he is dead and can no longer defend himself. He embraced this struggle with the greatest ardor because he believed that people were opposing the noble purpose of missions in general. He could not understand that individual members of this large group failed their purpose entirely by using the wrong methods.

We arrived at Mr. Bingham's and found in him the proud clergyman who is conscious of the fact that he exercises worldly as well as spiritual authority at the same time and, for that reason placing himself at the pinnacle, ignores the customary forms of social courtesy. Mr. Bingham invited us to visit his house as often as we desired and the mission doctor offered to accompany us on future excursions to the interior of the island. We declined both, however, partly in order not to incur any obligations to these gentlemen which would have cost us time and partly so that we might be able to inform ourselves about the condition of the island freely and without guidance from the missionaries.[†]

Kauike'aouli, the young ruler, returned to his palace on that same evening and sent immediately to Mr. Bingham in order to ask him for advice. General Miller had left the *Prinzess* before us and had used that afternoon to go riding. He had met Kauike-'aouli on his return to Honolulu and had himself introduced immediately. Kauike'aouli inquired at once about the gifts which the *Prinzess* was bringing him. He asked whether there was a

[†] Rev. Hiram Bingham and Dr. Gerrit Parmele Judd.

sword among them and was beside himself with joy when this was affirmed. As soon as Kauike'aouli returned from Mr. Bingham's, he sent a servant to tell us he would receive the letter from His Majesty the King of Prussia. Captain Wendt and I, accompanied by a North American merchant who was to serve as interpreter, then went to the home of Kauike'aouli.

It was a beautiful tropical night, the moon shone brightly and the dark-blue sky glimmered with many stars as the young ruler granted us his first audience. Two small, graceful Indian huts which belonged to the queen mother, the only surviving wife of Tamehamea [Kamehameha I], stood sideways on a wide open space in front of the royal residence. In front of these lounged several hundred Indians who were in the service of the royal family. In front of the door of one of these huts stood Kauike'aouli, and lying before him on fine mats were the old queen mother and four surviving widows of Riho Riho [Lihoiliho], the brother of the present ruler, who died in London. Kauike'aouli, who had been crowned king of the Sandwich Islands under the name of Kamehameha III, was seventeen years old and not especially big and strong. His face was so terribly scarred by smallpox and had become so extraordinarily swollen and copper colored, probably through the early and frequent consumption of strong, spicy, alcoholic beverages, that in our lands one could not easily find a man uglier than he. His eyes, his speech and his whole behavior during our stay on Oahu betrayed nothing of that through which his father had once attained such great power. Kauike'aouli was dressed in a white shirt, white pantaloons, a colorful vest and a white straw hat. He took off his hat when he received us and put the letter from His Majesty the King of Prussia, which Captain Wendt had the honor to present, into it. During our entire audience he remained standing in one and the same place. Although Kauike'aouli speaks some English, he had our entire conversation translated and asked at once about the gifts. When he heard that we had also brought gifts for his wife, in the event that he was married, he at once announced to those around him that he ought to marry very soon since even his friend, the King of Prussia, wished it. At the same time he asked us not to speak of the gifts, as this would cause jealousy among the ladies of his family.

During this conversation one of the servants who sat at the King's feet asked me to show him my large Peruvian hat of vicuña wool which I was holding in my hand. He immediately put the hat on his head, whereupon the Indians around him laughed heartily and teased him. In the hut behind the young ruler, several immense women lay stretched out on soft mats and showed great curiosity to see us.

We were then introduced to the queen mother, Queen Kaahumana [Ka'ahumanu] who sat kneeling all alone on a mat, and was wrapped in a colorful Chinese blanket which she opened only far enough for us to now and then catch a glimpse of her face. A white band covered her forehead. Seeing the enormous shape of Ka'ahumanu kneeling under the colorful blanket in the bright moonlight surprised us so, that at first we really didn't know what to make of it. She resembled an idol until she extended her hand to us in a friendly manner and, pointing to herself, repeated several times, "My Queen! My Queen!" Ka'ahumanu probably wanted it understood that she was the Queen and true ruler of the Sandwich Islands and not her step-son, Kauike'aouli, who, as long as she lived, merely assumed the title.

With this the first audience ended. It had been decided that the gifts should be publicly presented on the following day at the home of Kauike'aouli. At the same time we had received permission to explore the whole island for as long as we liked.

Gifts for Kauike'auoli
~ Two ~

June 25th. On the morning of this day we visited the market and also took the opportunity to carefully explore different areas of the city of Honolulu. We will have a detailed account of this later. It was not until about half past nine that the gifts were brought to land and transported on little two-wheeled carts to Kauike'aouli's home. He had assembled his court there in order to welcome us and to receive the gifts from His Majesty the King of Prussia in their presence.

As we entered the courtyard of the royal residence the guards, dressed in English seamen's uniforms, presented arms. Otherwise the soldiers of the Sandwich Islands are quite naked except for the malo and a piece of linen-like cloth which hangs from the shoulders. In the house of the King we found the great men of the kingdom assembled. They stood about, leaning against the walls of the room like statues. Kauike'aouli and John Adams, the Governor, sat on a bench and had us sit on the one opposite. Most of the foreign merchants of Oahu were also assembled for this festivity. The young ruler extended his hand to each one as he entered and often one could hear "Good morrow King! Good morrow King!" Kauike'aouli was dressed in white pantaloons, a black jacket trimmed with cord, a colorful scarf and a many-colored vest, but the awful figure of the Governor was in a blue dress-coat which was trimmed with some anchor buttons and whose tails almost touched the floor.

The home of the King is built in the style of the Indian huts but because of its size it is a veritable palace in comparison with them. Yet it is like a barn in comparison with the houses which some of the merchants and especially the missionaries have built in Honolulu. The building is about 140 feet long of which the first 120 feet form a single room down whose center run the pillars on which rest the rafters. The pillars in the center of the house, as well as those which form the walls, are round stems of the coconut palm. They are fitted with long cane which is interwoven with grasses and especially with the leaf stems of various delicate ferns. At the end of the house is an area separated by colorful curtains which in turn has two small rooms on each side and a larger room in the middle. The small rooms are for sleeping and dressing. In them lie large piles of fine mats, 15 to 20 on top of each other, arranged so that the one above is always finer than the one beneath. One can sleep extraordinarily well on these mats. Two pictures about three feet high and set in large gilt frames, one depicting the present king and the other the queen who died in London, as well as a third picture depicting the assembly of the Congress in Washington decorated the room between the two small rooms. The large room in which court was held had no ornaments. The floor was covered with fine mats and the furniture consisted of a large oval table of well-polished wood, two lacquered benches with backs, a table that stood off to the side on which water was kept, and several wooden chairs.

Right after our arrival the ladies of the royal family appeared. Ka'ahumanu, the old queen mother, led the way with measured steps. She was followed by the ladies Kinau, Kekauluohi and Kekau'onohi, all sisters-in-law of Kauike'aouli and the surviving wives of Liholiho, who had died in London. Also in the retinue were a niece of the deceased Prime Minister Kalaimoku, who had become well known by the name William Pitt

and Madame Boki, the wife of the shipwrecked Governor of
Oahu. She had accompanied King Liholiho to London. As they
entered, each of the ladies extended us her hand and Ka'ahumanu,
at her great age, made a very fine appearance. The ladies were
all dressed in very wide silk dresses, called missionary shirts,
which were drawn up at the throat by a cord. They wore black
silk stockings and shoes and their hair was very tastefully orna-
mented with the beautiful blossoms of the *Edwardsia chrysophylla*,[1] which was introduced from Tahiti. Ka'ahumanu wore
a straw hat trimmed with flowers and feathers which looked very
odd because of its age and shape. After the ladies had seated
themselves, some on chairs and some on the floor, and while the
servants who had been in the ladies' retinue took their places on
mats in the background, Kauike'aouli requested that the gifts be
presented.

Now the trunks containing the gifts were brought into the
room and opened in the presence of the assembled company.
Captain Wendt and I attempted to bring the things out in an
order that would produce the greatest effect. The assembly
showed great astonishment at the number of gifts but Kauike-
'aouli, sitting on the bench, exhibited such measured reserve at
first, that one soon had to take his behavior as an affectation.
The statues cast in iron, among which were those of Frederick II,
Alexander I, Napoleon, Blücher, etc., aroused the greatest joy.
The one of Frederick II was admired particularly and the King
had it brought to his seat in order to look at it more closely. The
military uniform, the plumed hat and especially the beautiful
sword seemed to please them exceptionally well. A very fine
saddle and harness was immediately put on a white horse and
aroused great joy. But most of all they liked the splendid paintings of His Majesty the King of Prussia and of Prince Blücher,
which Kauike'aouli had once desired to see. The pictures of the
various troops of the Royal Prussian Army which were also among
the gifts were passed around and around among the company and
were admired on all sides amidst the loudest comments. Among
the gifts which were meant for the assumed wife of Kauike'aouli
there was a very fine lady's hat trimmed with artificial flowers.

It especially aroused the curiosity of the young Queen Kinau who, in spite of her extraordinarily large size, still possesses quite special charms. Kinau had the hat put on her head and was admired by all. She also liked the jewelry especially well and wished to wear it. At this point we were most embarrassed to find that the bracelets and the necklace, although they had been made exceptionally large, would not fit. Only with the greatest effort and by squeezing the lady's neck considerably did we succeed in fastening the necklace—and she is, in comparision with the others, not at all heavy, but rather delicately built.

Kauike'aouli was asked to put on the uniform, which he did immediately in the next room with the help of his secretary, Halilei. But when suddenly the announcement "the missionaries are coming!" was heard, he quickly took it off again. When he returned to the room with the uniform and saw his sister-in-law Kinau with the jewelry, he immediately told her to take it off as it was not meant for her and that she would receive none of it. His sister-in-law obeyed immediately without any indication of anger or displeasure. The fine linens, the silken cloths, the toiletries and many other things aroused the envy of the ladies present because Kauike'aouli kept everything for himself.

During the whole time that the gifts were being presented, Ka'ahumanu, the queen mother, sat quietly and sadly. She could hardly conceal her envy and preferred to appear ill. Two servants stood beside her and had to constantly fan fresh air towards her. So much did the old woman desire a cane with a harmonica, which we had given to John Adams, the governor, that she seized it and immediately experimented with her musical talents—right there in the midst of this high assembly.

After our task had been accomplished we took leave of the assembled court. It was a very hot day on which we presented the gifts and since we had been occupied with this business for about four hours we were extraordinarily thirsty. Some of the foreign merchants who resided there indicated to the King that he should offer us something to drink but he replied that the missionaries had forbidden it.

The gifts made a strong impression on Kauike'aouli and on

all the other great personages of the kingdom. Even though the former had behaved throughout in a very aloof and artificial manner, a manner which had obviously been dictated by the missionaries, he had nevertheless told some of the English merchants that he was quite ashamed because he had given His Majesty the King of Prussia only a feather cloak—such a trifle—and now received so extraordinarily many things in return, which he did not know how to repay. It is in fact true that, as often as the English sent gifts to the Sandwich Islands, these never exceeded the value of those which we had the high honor to present.

The occasion for these gifts which His Majesty Our King sent to the ruler of the Sandwich Islands stems from the first visit of the Royal Prussian Maritime Ship *Prinzess Louise* to Honolulu. Kauike'aouli, then much younger, had heard much about the deeds of the Prussian Nation in the great war of liberation against Napoleon and that the greatest part of its happy success had to be credited to Prussia alone. He had been told much about Prince Blücher, our field marshal at that time, and a certain admiration of the great deeds of this man had taken possession of him. He often expressed the wish to at least be able to see a picture of this brave man. In his admiration for Prussia he sent His Majesty, the King of Prussia, a many-colored feather cloak, along with a letter explaining the great value of this gift, as this cloak had once been worn by Kamehameha I in the battles which resulted in the subjugation of all the Sandwich Islands under his rule. This feather cloak as well as the letter which accompanied it have been put into the custody of the Royal Museum in Berlin.

Upon receiving this gift, His Majesty the King graciously took the first opportunity to send Kauike'aouli the picture of Prince Blücher along with the many other gifts. This opportunity presented itself at the return voyage of the *Prinzess Louise*.

To Madame Boki's Home
∼ Three ∼

Since we anticipated our stay on Oahu to be very short we lost no further time. Several hours after the festivities and the presentation of the gifts were over we busied ourselves with preparations for a trip into the mountains. We set out about five o'clock in the afternoon. Dr. Rooke, a most amiable Englishman who practices medicine in Honolulu, was so kind as to accompany us on this as well as all following excursions. Kauike'aouli had offered us his servants and we used a great number of them to carry the baggage, the instruments, the muskets and the food. Most of them went quite naked except for the malo and were not very friendly on account of their heavy burdens. Only a few mestizos who were also among the borrowed royal servants wore fairly complete clothing and also assumed command of the simpler Indians. It did not take us long to realize that travelling on the Sandwich Islands is much more unpleasant than in the Cordilleras of Chile. There are no pack animals here and all baggage must be transported on the backs of men. Besides that, the ordinary foods in this land are of a type which takes up much room and yet gives little nourishment. As it turns out, a man eats almost as much in the course of a day as he has carried and if one does not take everything along it could happen that one could be in the interior of the island for several days without finding anything to eat. It was Saturday when we departed from Honolulu and since no church services are conducted on that evening the inhabitants of the city use it for entertainment. We were most astonished when we suddenly saw women on horseback in all the streets of Honolulu, mounted like Amazons and racing by at full gallop. It was a truly comical sight to see these women, especially those immense figures of the royal family, on horseback. They sit in the saddle like men and have only a little tapa wrapped around their legs. Without hats, with garlands in their hair and a riding whip in their hand they ride along at a steady gallop, usually alone, but sometimes in the company of several men and women.

We had to climb several miles up into the mountains on this day in order to reach a dwelling where we could find good quarters, so we did not spend much time in the city but quickly undertook our journey. We had scarcely left the gardens of the city houses—which are for the most part planted with beautiful flowers *—when we came upon expansive plantations of *Arum macrorrhizon.*[2] These are known here by the name of tarro [taro] fields. What a sight for us to see such great fields of this precious food-plant! Taro is planted in water, for which purpose they have dug huge square fields to a depth of 2 to 3 feet and filled them with water. The borders of these basins, which divide the property of various owners, are also—at least in densely cultivated areas—used as foot-paths and are densely planted with banana trees. Near the water and in this fertile ground they attain a giant size. Right next to these fields lie fields planted with sugar cane, which is used here only for food. The bluish green of the sugar cane is a remarkably beautiful contrast to the light green of the young banana leaves and the velvety color of the taro leaves. How much more beautiful is the sight of these tropical plants in their own land! We get to know them only as stunted specimens in

Hibiscus tiliaceus, Cordia sebestena, Canna flava, Edwardsia chrysophylla, and others.

our greenhouses. Everywhere along our way we found the *Lythrum maritimum Kunth*, the *Hydrocotyle interrupta D.C.*,[3] *Jussiaea augustifolia*,[4] and others, and in the water basins of the taro patches, especially in those which were uncultivated, grew a great deal of *Potamogeton*[5] and *Chara*, which were covered with countless little snails. Occasionally our way took us through shallow, wide ditches in which the Indian women were bathing. If anyone tried to disturb them, they splashed vigorously.

The valley of Honolulu, in which we now travelled to the northeast, runs along the whole southern coast of Oahu and is nothing more than the continuation of this flat coast which is bordered everywhere on the north by the high mountains. At Honolulu this valley is wider than usual, indeed in some places the distance between the mountains and the shore is an hour or more. In all areas that are richly supplied by water, as for example immediately behind the city, the fields of this valley are well cultivated, but further east the plains lack water and are used as pastures for grazing. The mountain chain which extends across the isle of Oahu from southeast to northwest and whose individual peaks rise to over 3000 feet has numerous transversal valleys which run almost in a straight line down from the ridge and open to the south. Three of these transversal valleys open at Honolulu and we traversed the entire length of each. The westernmost of these valleys is the most famous and was called the Kuaroa Valley.[†] Here Kamehameha I fought the famous battle which made him sole ruler of the Sandwich Islands. He drove the enemy into the narrow passage of this valley and many hundreds met death in their speedy retreat.

The view that one enjoys from this valley of the plain of Honolulu and the harbor with its roadstead is indeed one of the loveliest that we experienced. It is therefore all the more surprising that the view of this area which serves as the frontispiece of the London edition of Stewart's *Journal of a Residence in the Sandwich Islands* [††] is one of the worst pictures with which recent

†Note: Meyen refers to the Nuuanu Valley as Kuaroa. He was misinformed.
††Stewart, Charles Samuel, 1795-1870.

travel descriptions have been adorned. The Kuaroa [Nuuanu] Valley rises very gradually, its direction being north 15° east with a 10°eastern declination. At almost every step something caught our eye and the narrow path between the taro fields also demanded our attention, if we did not want to fall into the deep mud. A great number of small water plants were to be seen everywhere but unfortunately there was no time to examine them, else we would not have made much progress and would only have collected a few specimens.

In the middle of the valley flows a small river which comes down from the mountains. Its water is channeled off right and left to the taro fields. In this region the taro fields are constructed just like the rice paddies in China, that is, in such a way that the water coming down from above flows through the fields and can thereby be drained off from one basin into the next.

The area was very lively on this evening and everywhere we observed great activity. Busily the Indians hurried past, never forgetting their greeting: "Aloha!" On their shoulders they carried pretty, very colorful calabashes containing water which they had drawn further up in the valley and would sell to the wealthier people in Honolulu, since the water near the city is not particularly good, indeed has a foul taste. Other Indians carried wood which they had cut in the mountain forests, still others the bark of that famous and useful plant from which the fine fabrics (tapas) as well as nets and rope are made. Still others carried home taro in big bundles in order to prepare it on that same evening for the following day, the general Sabbath. The big tubers of the *Arum* they tie together with its leaves into bundles of 10 to 12, hang one such bundle at each end of a short pole and carry them away. This manner of carrying things is quite common in the Sandwich Islands; be it food, wood, water or anything else, it is carried hanging from the ends of a short pole across the bare shoulders.

All of these Indians who were returning from work were quite naked except for the malo. They greeted us in a friendly manner and in their talkativeness they could not keep from beginning conversations with our attendants. Our attendants were also very talkative and extremely merry. It was new to them, but very enjoyable, to accompany a man who had set out to collect stones, plants, insects and other things of that sort whose purpose they did not know and therefore had all the more reason to make fun of. The Indians travelled very well but could not remain silent for even one moment. As they walked they sang monotonous songs until one of them made a joke—often about the most trivial thing—which was followed by considerable laughter.

After the sun had set, the temperature of the air was more pleasant, especially since we had climbed to several hundred feet above the plain of Honolulu. Unfortunately the mountain tops were covered with thick clouds which threatened us with rain.

As soon as we reached the elevation where this valley in which we journeyed is surrounded on both sides by porous basalt, the grade was no longer so gentle but often rose sharply. Much of the rock wall on both sides of the valley is formed by sheer cliffs in which one can plainly see the stratification of the rock.[*] In these areas which are bare of all vegetation, the basalt,[**] which contains olivine, is often peculiarly formed—similar to some of the strange rock formations in our own regions. Generally, however, the grade of these stone walls is not so steep and then they are completely covered with the lush vegetation which is native to these tropical regions. In the valley itself, which is about a half mile wide, basalt boulders which have fallen from the mountains lie about. Usually these were broken off through the intrusion of vegetation. On that same evening when it had already gotten dark and we were close to our goal, such a rock slide took place on the western side of the valley. The falling rock made such a terrible noise, that for a time we were in doubt as to its cause. It resembled the muffled roar of a volcano such as one can frequently hear in Chile.

[*] Basalt somewhat blistered with small augite and olivine crystals mixed in. It is weathered on the surface but of a beautiful grayish-black color in the interior. Augite and olivine, which resist weathering better, protrude from the surface.

[**] Porous, scoriaceous basalt full of small round or larger oblong hollows; the color of the solid mass is gray but the walls of the hollows are covered with a brown coating. Small crystals of olivine can be seen here and there enclosed in the basalt.

From here on the valley of Honolulu rose ever more steeply and as the cultivation of the land became less frequent with increasing elevation, the wild plants became more prominent. The fields of taro, which are planted under water, extend in this valley to an elevation of 800 feet and give it a very interesting appearance when seen from above. These fields are generally square pieces of land, 40 to 50 feet wide and just as long. Since it is a sloping valley they are laid out in terraces so that, as has already been mentioned, the water from one basin can drain down into the next when it has attained the necessary height in the former. The leaves of this precious plant rise only slightly above the basin and the individual tubers are planted farther apart than our potatoes—somewhat like cabbages—but also in straight rows. Just as among our cultivated plants, those which are grown primarily for their roots seldom bear flowers and fruits; so it is here with the taro plant, the *Arum macrorrhizon.*[6] We have found only three specimens of this plant in bloom and these stood by a little stream right next to the big fields and had become wild. Right next to them stood a blooming specimen of *Caladium esculentum,*[7] which is probably also cultivated here on Oahu. At an elevation of 800 feet the cultivation of the so-called dry taro begins. This plant is the same *Arum macrorrhizon*[8] which is planted down in the plains under water. It is also planted in very good soil but the tuber attains neither the size nor the good taste of the varity which is cultivated in the water. It is not used for the preparation of poi, the favorite dish of the Sandwich Islanders, which we will discuss later.

Where the cultivation of the wet taro ceases so does that of the banana, which we did not see planted above an elevation of 800 feet on Oahu. But we did find wild bananas in the interior of the forests for 400 to 500 feet farther up. These, like the *Musa textilis* in Manila, are used for various fabrics.

The sweet potato * is also a much cultivated plant in this valley. Little care is expended upon it, however, so that it is quite bad and not to be compared with the Peruvian one. It seemed to us that the sweet potato is just simply not very tasty in damp, tropical regions. In Brazil as in China, on the coast of

Convolvulus batatas.[9]

Manila as on the Sandwich Islands, it is a soft, watery tuber to which the ordinary potato is to be preferred. However, the sweet potato which is cultivated on the plateaus of southern Peru—for example in the valley of Arequipa—is to be preferred above all other vegetables.

From the number of plants which we collected on this route we will point out a few which are especially important to the inhabitants of this area. In some of the drier areas the *Tephrosia piscatoria Pers.*, which is one and the same with the *Tephrosia toxicaria Gaudich.*,[10] grew abundantly. It is one of the few plants in the legume family which contain a strong narcotic element, for which reason the inhabitants of the Sandwich Islands use it for catching fish, just as we use the berries of the *Cocculus.* Among the great number of *Convolvulaceae* which spread along the sides of our path, we found the *Ipomoea bona nox*[11] and *Ipomoea cataracta End.*,[12] whose roots the natives use as a very strong purgative. It would be desirable for the doctors there to pay special attention to this plant.

Frequently the fields were covered with a *Zingiber*,[13] whose roots resemble the East Indian ginger, but have a bitter taste. The plant is not yet cultivated here but grows wild in great abundance. The natives gather the roots, salt them and then eat them. Another *Scitamineae*,* also grows here in abundance. Its roots are used as an alternate *Curcuma* for turmeric, that famous East Indian dish which the English like to eat and have on their table every day. A little *Solanum*[15] with black berries and similar to our *S. nigrum* also grows here not infrequently and is often eaten. The berries of the plant taste quite good.

The little river which comes down from the mountains and flows through the length of the valley has very high banks further up. These at times make a most pleasing impression because of their lush vegetation and especially because of the beautiful combination of different varieties of plants. The stems of the *Musa*,[16] which grows wild in such places, often reach a height of 20 feet and more and are also relatively thick. From the branches of the tall trees descends *Ipomoea bona nox*[17] with its large white blossoms and great, heart-shaped, shining leaves, resem-

Curcuma longa. [14]

bling the *lianas* in the jungles of Brazil. The dense foliage of other trees is thickly covered with *Ipomoea palmata Vahl,* [18] whose flowers resemble those of *Ipomoea variabilis*. One of the greatest beauties of this vegetation is the magnificent *Jambosa malaccensis D.C.*[19] with its heavenly rose-apples which the Indians call ohias. As long as the fruits of this beautiful myrtle tree are not yet ripe they hang like eggs from the branches and are a gleaming white. As they ripen they redden until they finally become cherry red and then they have a very pleasant, slightly acid taste. They quench the thirst and are, even if enjoyed in large quantities, very easy to digest. One finds the fruit very frequently in the market of Honolulu and much of it is consumed in the huts of the Indians.

The higher we climbed the more lush and diversified the vegetation became. The fields consisted of a number of very interesting *Cyperaceae* and grasses. We collected there the *Cladium leptostachyum n. sp.,* the *Rhynchospora castanea n. sp.,* [20] the *Cyperus auriculatus n. sp.,* [21] the *Cyperus owahuensis n. sp.,* [22] and the *Panicum pruriens,* [23] which we found here as well as in many other places on the island. All of the little shrublike plants which grow here are covered over and over with the most diverse *Convolvulaceae* and a host of other climbing plants, especially the *Cardiospermum halicacabum L.,* make the thicket all the more impenetrable.

We had been travelling in the dark for quite a while and the steady rain, which had begun shortly after sunset, was becoming extremely unpleasant when we finally reached the building which was to serve as our quarters for the night. The house belonged to Madame Boki, who had placed it at our disposal for the night. It was used by Madame Boki as well as by the royal family as a winter palace to which they could withdraw when it became too hot in the plains. The temperature here was most pleasant even though the house stood only between 600 and 700 feet above sea level. During our stay there we never measured an air temperature over $17°R$ ($21°C$, $70°F$). From this house one has an extraordinarily interesting view. The entire valley which lies behind the city of Honolulu is covered by the most lush vegetation and the liveliest green. From here it falls off gradually to the ocean, where the shores are adorned on the one side by coconut groves, and on the other by the many hundreds of huts and houses of the city as well as by the retaining walls of the royal fish ponds. On both sides of the valley rise high and steep precipices, often exceeding a height of 1000 feet, which are also covered by beautiful plants and even display little waterfalls.

The house in which we stayed was very roomy and built quite like other Indian huts. A partition of matting divided the bedroom from the living room. Besides a few mats for sleeping there was nothing else in this roomy building but a lamp, for it is customary on the Sandwich Islands to have a light burning throughout the night. For this purpose they use the shells and the oil of the nut of *Aleurites triloba,* [24] which grows here everywhere in the forests in great numbers. In addition we found in the house a very thick wooden bowl, 6 feet long and 2½ feet wide, which is used for the preparation of poë [poi] , the mashed pulp of the taro root. A bowl of this type, smaller or larger, depending on the need, is certainly not lacking here in any household. The only other article which was to be found in this big house was a shallow calabash for washing.

Madame Boki had instructed the inhabitants of the little huts which stood in the vicinity of this big country house to receive our people well. As a result there appeared on that same evening, soon after our arrival, a very large calabash of poi, which was devoured by our people with a mighty appetite. The preparation of taro, the basic food of the inhabitants of the Sandwich Islands, is extraordinarily diverse. The roots are generally between the size of two fists and the size of a child's head. They are baked in the earth and then eaten with or without salt, like bread. The tubers are also cut into slices and fried in fat, or, what is most common, they are boiled, then kneaded in large wooden troughs— which we saw up to 10 feet long and 3½ feet wide—by beating them with large stones and adding a bit of water. Dry taro is prepared and eaten this way. To the mashed pulp of the wet taro they add more water and leave the mass to ferment, which ordinarily takes place in only 24 hours. This semi-fluid pulp, called *poi*, is the favorite food of the Sandwich Islanders and the enormous quantities that they can consume are often unbelievable. Since the use of spoons has not yet been introduced into this land, the Indians must eat this pulp with their fingers, which looks somewhat shocking. Generally the whole company sits around a bowl of this precious pulp, then each sticks his first two fingers into it, swishes them back and forth a few times so that as much as possible clings to them, then brings his fingers with an especially artful movement to his mouth, inserting half the hand into it in the process, and licks it clean with his tongue.

After my people had consumed their meal, which had tasted especially good to them, they stretched out in all directions on the mats but chattered so incessantly, that we were forced to order them either to be silent or to leave the house, whereupon all were quiet and spoke not another word during the whole night—something which was undoubtedly most difficult for them.

The garden in whose midst our house stood was surrounded by a hedge of theti* [ti] which sometimes displays green, sometimes blood-red leaves, indeed, one often sees both colors on the same plant. Within this enclosure stood several trees of the *Acacia heterophylla* Wild., [26] which deserves attention for several reasons. Its name comes from the various shapes of its leaves which are usually simple and sword-shaped, but on some branches are pinnate and even double-pinnate, which has a most peculiar effect when occurring on one and the same tree. What also adds to this effect is the lighter green color of the little, pinnate leaves in contrast to the large, sword-shaped ones. Furthermore this tree, whose height and thickness is often quite extraordinary, must be regarded as the representative in the higher northern latitudes of the large genus of *Acacia*, for no variety of *Acacia*, whose original home is actually New Holland [Australia], can be found further north.

Dracaena terminalis Jacq.[25]

Nuuanu and the Pali
~Four~

A heavy rain which fell on the morning of the following day soaked us so thoroughly, that we soon had to seek shelter in our house again. We wanted a warm breakfast because the temperature 16.8°R (21°C, 70°F) was very unpleasant. We were surprised, however, when the people explained that since today was Sunday, the enjoyment of all warm food was prohibited by a tabu, upon which the missionaries insisted with all their power. Nothing could have struck us as more ridiculous here in this charmingly beautiful spot of nature. I immediately took the wood, kindled the fire myself and made my coffee. As soon as the Indians saw the fire burning they shouted with joy, fetched wood and didn't give the tabu another thought. Instead they repeatedly uttered abuses against the missionaries, especially since they had only dry taro to eat on this day, which didn't taste as good to them as their poi.

The observance of Sunday as it has been introduced to the Sandwich Islands by the missionaries would be well recommended for convicts in a public reformatory but not for such good-natured and poor people as the inhabitants of the Sand-

wich Islands. On this day all pleasures are denied until sundown
and the people must go to church morning and afternoon. Even
taking a walk or riding is forbidden and in recent times they have
also enforced this order against the foreigners with the greatest
strictness, taking away their horses on such days and sentencing
them to fines of 100 piasters. Our friend, Captain Wendt, wanted
to please us by coming on horseback to join us that day in order
to continue the journey with us. Mr. Wendt sent to the Governor
and requested permission to use a horse on this day, which could
well have been granted him as a visitor. Nevertheless, the request
was denied. The enjoyment of warm food, as the lighting of fire
in general, is completely prohibited on Sunday. This law falls
most heavily upon the poor Indian, whose choice of foods is
very limited. The rich can well endure cold foods, as these
generally tend to be much finer. Just how important the enjoyment of food was to these religious men, the missionaries, on
the holy fast day one can best learn by perusing the journal of
the famous missionary, Stewart. In this book it seldom fails
that on Friday or Saturday the author speaks of all the work
and inconvenience the wives of the missionaries assumed in order
to prepare a sufficient quantity of food for the following day,
holy Sunday.

[June 26] About seven o'clock in the morning the rain stopped and we immediately set out on an expedition into the surrounding area. Accompanied by a few of my people, we entered
the thicket and tried to reach the steep cliffs bounding the valley
on the west. We had only gone a few steps from the house, however, when it became almost impossible to proceed. The immense
mass of high, bush-like ferns, of *Pandanus* and *Scitamineae* [27] is
so thickly woven together through countless *Convolvulaceae*,
that one must first break through all the vines in order to clear a
way. Further on, where the vegetation becomes more arboreous,
the *Pandanus* and *Bromeliaceae* occur in the greatest number.
Some of them climb the trees and enclose them, often with
hundreds of branches, so that their foliage becomes impenetrable.
Here three or four Indians ran ahead and, by together throwing

themselves on the seven to eight foot high cover of vegetation, pressed it down far enough so that one could walk on it if necessary, since it was simply impossible to hack through the mass and clear a way. So it sometimes happened in this area, that we quite unawares wandered about on a cover of vines, often ten feet above the ground. When we had to climb down from this mountain of vegetation at some ravine or other in order to get onto the next one, we were able to get an even better overview of the enormous mass of vegetation which is present here on very small stretches of land. Very soon a great number of trees became apparent, as well as a very great diversity in the genera and species to which they belonged. The trees of these forests do not attain the immense height and impressive breadth dimension which make the forests of Brazil and India so picturesque. On the other hand there is no doubt that the forests of the Sandwich Islands have far more underbrush and are likewise much richer in herbaceous plants. Here we found the plant called *mamaku* [mamaki], from which the Indians make their ordinary tapa. It belongs to the *Urticaceae*[28] and is the *Neraudia melastomaefolia Gaud.* A new species of this genus* grows here very plentifully and is likewise used for making the cloth. The *Böhmeria albida Hook.*[29] is actually the plant from which the finest tapas are made and it also commonly grows here in the forests and is widely cultivated in other parts of the island. It is called "kuku."[†] The making of this cloth from the fibers of the young bark of this plant has become well known through earlier travellers and we therefore refer the reader to Cook's *Voyages.*

The Indians especially pointed out to us the high, beautiful trees of the *Aleurites triloba.*[30] They are called kukai [kukui] and are used for many things. The shell and the oil of the fruit are used for burning and the bark is processed for the tanning of leather. The stems of this plant are frequently covered with large lichens, of which the *leka*** is eaten by the poor people. In addition we also collected here *Parmelia perforata var. melanoleuca*

Neraudia glabra n. sp. N. foliis late ovatis acuminatis crenatis utrique glabris.
**Parmelia perforata Ach.*
†kuku — this means "to beat" as in the making of tapa.

and var. *ulophylla, Usnea australis Fr., Sticta lurida n. sp.* and several *Jungermanniaceae* and feather mosses, which have been described as new in the botanical portion of the account. Small *Jungermanniaceae* grow here on the thallus of the lichens. The shoots of the many shrub-like ferns which grow here are also eaten both raw and cooked by the poorer classes of people and many of them taste quite good. These contain an extraordinary amount of starch, which occurs in the cells of several species in larger grains than we have seen in any other plant, even in the *Cycadeae*. Indeed, the good Kanakas, as the Sandwich Islanders call themselves, disdain very few plants. On our excursion it was very common for the people to point out some plant to us and say "kau-kau," which meant that the plant was edible.

The farther we tried to penetrate towards the rock wall, the denser the vegetation became, and the little drop-offs, which we chanced upon on this side excursion, could only be crossed with the greatest difficulty and risk because of this dense vegetation. From time to time we stood still in order to take in and impress upon our memory the fantastic pictures which the lush vegetation presented to our eyes. Thick tree trunks entwined with *Dracaena, Pandanus, Convolvulus,* among others, and adorned with colorful lichens, together with magnificent ferns, presented the most beautiful sight any travelling botanist could wish. The largest variety of the *Asplenium nidus,* whose enormous leaves are two to three feet long and relatively broad, was found growing beside small spear-shaped leaves of a *Pteris* species, with *Piperaceae** in great numbers, pretty little *Jungermanniaceae,* feather mosses, etc., and all on the same tree. What a sight! The curiously shaped *Charpentiera obovata Gaud.* carelessly hangs its blossom clusters down over the *Lobeliaceae,* of which so many new forms became familiar to us through Mr. Gaudichaud.[†] Besides the *Cyrtandra cordifolia G., C. grandiflora G.,* and *C. lessoniana G.* we also found here a new species of this beautiful

---

*Peperomia verticillata,[31] *P. leptostachys Hook., P. membranacea Hook.*

†Gaudichaud—Beaupre, Charles, 1789-1854, was a French botanist who made two trips to the Hawaiian Islands, one in 1819 and the other in 1836.

genus.* In what unbelievable numbers we found the shrub-like ferns growing here! —such as the *Blechnum fontanesianum G.*,[33] *Aspidium exaltatum*,[34] *Polypodium pellucidum*,[35] among others, and beside them the many *Prasium*[†] species and the *Plectranthus parviflorus*. The only plants we found missing here were orchids and *Umbelliferae*. Aside from the *Hydrocotyle interrupta*,[36] nothing of the latter family could be found. The thick vegetation finally presented a limit to our progress and we did not reach the cliffs where we certainly would have found many interesting things. In this thicket we frequently came upon a large, bright green spider which placed its egg sac in the center of its web. We have brought this spider along in alcohol. The number of insects we collected here, as indeed everywhere on Oahu, was extremely small. Only under large stones did we find some small beetles but under the bark of the trees we saw only millepedes and some spiders. The entire collection of insects which we brought along from Oahu consisted of thirteen specimens—so great is the scarcity of insect life here.

In the span of several hours we collected an extraordinarily large number of plants. The Indians were very helpful to us in preserving them—as though they had often watched this kind of work. We had not received such help from the people either on our travels in Peru or in Chile. There the people usually sat around us making fun of us for having to occupy ourselves with such things. They lacked any talent, however, to be of help.

As soon as our collected treasures had been packed, we again set out on our journey. We travelled the length of the valley in order to reach the mountain ridge. Everywhere we were accompanied by the splendor and beauty of the vegetation. Our atten-

---

Cyrtandra ruckiana n. sp.[32] *C. foliis ovatis obtusiusculis subtus integerrimis, supra puberulis, subtus ferrugineo pubescentibus, pedunculis unifloris ebracteatis, bracteis late ovatis obtusis, calycis tubulosi pubescentis dentibus actiusculis.*

†Note: Previous to the work of Meyen, Charles Gaudichaud had collected four members of the Mint family in the Sandwich Islands. Since they had a fleshy fruit he put them in the genus *Prasium* which has a single species in southern Europe and southern Asia. More recent studies have proven that the Hawaiian species are really in a local endemic genus and they are all now called *Phyllostegia*.

tion was also drawn to several small but very quaint Indian huts. Only very small pieces of land in the vicinity of these huts were cultivated. All the rest which could bear fruit for thousands and thousands of people is still completely wild and the dense vegetation presents a barrier to any penetration. Gradually we began to see more and more of the *Metrosideros polymorpha Gaud.*,[37] one of the most beautiful flowers whose brilliant scarlet delights the eye. The Indian women also love this flower very much. They fashion thick wreaths from the blossoms of this tree, which they wear on their heads, and on this day they also presented us with one such wreath as a token of special honor. A little bird, the *Nectarina flava,*[†] lives primarily on the pollen of this plant and can always be found in the vicinity of such trees. The little children, some only three years old, catch this pretty yellow bird— which is even smaller than our wren—by brushing a kind of birdlime on the blossoming branches. A sling is fastened to one foot of these poor little animals and they are then tied to the malo —that piece of material which they have wrapped around their hips. The children run around, often with several of the little birds hanging from their bodies. These remain quite still and do not struggle at all.

Everywhere the *Dracaena terminalis*[38] was used in great quantities as hedges around the small huts. The root of this plant has an extremely high sugar content. They have tried to distill a rum-like brandy from it, which turned out to be quite good. However, the expansion of this new industry—just as the establishment of any factories and plantations—is forbidden. When one has chopped off the tuberous root of this plant one can stick the stem back into the ground and it will root anew. According to Solander[††] (manuscript), six varieties of this plant

---

†The bird is probably the 'amakihi now known as *Loxops virens.* Wilson and Evans (1890-99; *Aves Hawaiiensis)* list the name in the synonymy for *Loxops virens virens* of the Big Island. The source of the name was Bloxam's *Voyage of the Blonde* (1826) but the exact locality from which the bird came, whether Oahu or Hawaii, was uncertain. Meyen probably had access to the *Blonde* report and used the same name for the bird. Bloxam gives the name as **amakee.**

††Solander, Daniel Charles, 1736-1782. He was the doctor on Capt. Cook's first voyage.

are supposedly cultivated on Tahiti, each having a special name,
three of them being red and three white. When we reached an
altitude of 1200 feet the physiognomy of the vegetation changed
very noticeably. The *Musa*, as all *Scitamineae*—of which the
Canna indica var. flava accompanied us the farthest—disappeared
and little bushes as well as a great number of *Peperomia* species
appeared. Here we collected *Peperomia verticillata Sp.*[39] and
Peperomia tetraphylla Hook. Plantago quelcana,[40] *Oxalis repens
Thumb.* and a new *Atriplex* * also grew here. Our way rose very
steeply, now going up, now down again and everywhere boulders
of porous stone lay in our path. The valley itself was still enclosed
on both sides by cliffs which rose almost perpendicularly and
were covered with trees and bushes. Finally we reached the ridge
of the mountain range and were surprised to suddenly find ourselves at the edge of an immense slope which stretched down to
the level of the sea. Here we enjoyed one of the most beautiful
views to the north and the south. The latter would have been
even more beautiful if the valley through which we had climbed
did not have a bend in its last quarter, thereby obstructing the
view to the valley of Honolulu. In general, the whole mountain
range, as well as all its summits, falls off very suddenly. Great
stretches of these slopes are bare, devoid of all vegetation. The
mountain range is called Pele.[†] Its slope has a horseshoe-shaped
configuration below whose western point—called   Kuaroa
[Kualoa] —lies N. 10°W. Right next to it, in fact N. 8°W., lies the
small rocky island of Makoli [Mokulii]. Behind it is situated a
small anchorage which is enclosed to the south-east by a prom-
ontory. After the death of Cook, the English explorer ships
under Clerke and King sailed by the same bay and did not dare
to enter because of the stormy weather.**

    The promontory of Mokapu with a little village on the beach
lies N. 15°E. The shallow water of the ocean around this village

Atriplex oahuensis n. sp.[41]

    [†]Meyen must have meant *pali* which in the Hawaiian language refers to the
sheer cliffs and not to the mountain range.

is enclosed by the walls of a coral reef which resemble those enclosing the royal fish-ponds at Honolulu. The basins here primarily serve the purpose of catching sharks which can swim in through small openings but can't get out again. The kind which is caught here is supposed to be about three feet in length and is supposedly eaten with relish. The easternmost point of the northern coast which we could see from our position on the slope of the mountain range lay N. 14°E. Down in the Koolau Valley, which lay right at our feet, we saw the remains of the ridge of an old crater. The ridge of this crater does not rise up far enough to even have its own name in the Kanaka language. Even from above, one could observe that the vegetation in this valley was not as lush as in the valley on the southern side of the island. The reason which is given for this is that it very seldom rains there—which is probably quite correct. The trade winds blow across the northern side of the island and, through the influence of the land, change daily and hourly on the southern side. Regions in which the trade winds blow with force have no storms and generally no rainfall. The northern coast of the island of Oahu seems to have the same climate as such regions. We were most struck by the extraordinarily strong wind on the ridge of the mountains. It was stronger than anywhere else on our travels— even stronger than what we experienced on the plateau of Tacora —and with all that, the surface of the ocean right off the northern coast was completely calm while on the southern side it was agitated by a strong ocean breeze. As we climbed down from the heights all wind ceased and down below there was the calmest, most beautiful weather. It seems that the north-east trade winds leave the surface of the ocean at some distance from the northern coast, move diagonally up to the ridge of the mountains, blow across them and continue on out. Mr. Alexander von Humbolt also called our attention to such an extraordinarily strong wind which is supposed to blow constantly at the peak on Teneriffe. Other travellers found such a wind at the peak of Mt. Aetna. The explanation of this phenomenon is probably not, in our opinion, so very obscure.

Approaching the island of Oahu from the south, one notices in the mountain range on the left side of the transversal valley—whose northern slope we had now reached—a mountain which is especially high and whose peak is flat. From a distance, looking at it through field glasses, we thought it to be a volcano. Now, however, that we stood very close to it, we realized that it was cut off vertically on the north and formed only a wall. The peak of this mountain is by no means the highest on the island but rather a summit in the eastern portion of the range. Recently it was supposed to have been measured barometrically by Mr. Douglas, an English astronomer. Before we descended into the northern valley we set up the barometer in order to measure the height of the mountain ridge. Nowhere could we find a means of securing the instrument on the mountain. We had to climb down its northern slope and secure the instrument about 20 feet below the highest point on a root which protruded from a crevice. Here the barometer stood at 27.25 British inches at an air temperature of 18.1°R (22.6°C, 72.7°F) and a mercury temperature of 18.6°R (23.3°C, 73.9°F). At Honolulu—20 feet above sea level—the barometer stood at 30.10 inches at an air temperature of 20°R (25°C, 77°F) and a mercury temperature of 19.8°R (24.8°C, 76.6°F). Accordingly, the height of the mountain ridge is equal to 2821 Prussian feet.†

When this task was completed, we climbed down the steep slope of the mountain range. The crest of the mountains consisted of a pumice stone conglomerate* and somewhat deeper there was bedrock of true pumice stone with small and some large pores. Here we saw how the volcanic rock crumbles into a red soil, which occurred here and there in layers of varying thickness. A number of Indians from the neighboring huts—men and women, as well as children—had assembled and merrily followed us for their own entertainment.

Never again, neither in the mountains of the old world nor in those of the new, did we see a descent like the one we had to

---

*Pumice stone conglomerate; pieces of reddish-brown pumice stone are held together by a brownish-red and grayish-black jasper-like mass.

†2906 feet. There are 12.36 American inches in a Prussian foot.

negotiate here down to 1200 feet. There were only small ledges on which one had to climb up and down on the slopes of this hard, smooth volcanic rock. At one part of our descent our path took us past several regular columns of black basalt.[*] Its fracture surface is gray and also contains olivine. The numbers of Indians who made their way down with us—men, women and children, all in a gay confusion—gave the surroundings a liveliness worthy of being depicted by a talented artist. All of the Indians were quite naked except for the malo or some small piece of tapa.

Our descent on this bare rock wall which was so completely exposed to the effects of the sun progressed most laboriously and slowly. We were all tormented by severe thirst but were sustained by the hopes of a spring which was supposed to be at the foot of the mountain near our way. When we reached it, however, we found it buried. A huge mass of rocks had dislodged on the slope and covered it forever. Not until we had descended far down into the plain did we come upon a spring which supplied us with the necessary water, even though it was quite foul. The water of this little spring was filled with *Conferva* and *Chara,* of which we found two interesting species here.[**] One of these *Chara* species belonged with its thread-like stem to the old genus *Chara,* and with its delicate little branches to Agardh's[†] new genus *Nitella.* From this one can judge the correctness of the genus character-istics on the basis of which Mr. Agardh presented the world with such a large number of algal genera. Unfortunately these were accepted by most algal taxonomists, who almost never concern themselves with the physiological investigations of these things. It is for this reason also that matters in this field of botany are getting worse every day. The types of *Chara* which are classified under *Nitella* do not even present a logically correct and clearly defined subdivision of the genus *Chara* — and they even make a genus out of them! Among the *Chara* just mentioned, it is not seldom the case that some limbs of the branches have single and

[*]Basalt-gray and dense, interpersed with indistinct little white crystals, a feldspar -like substance (Labrador?) and individual embedded olivine.

[**]*Chara armata nob.* var. *diaphana.*

[†] Jacob Georg Agardh, Professor of Botany, Lund, Sweden.

some double skins. Another species of this genus* was so plenti-
ful in this water that it filled the entire space of the basin.

The spring of which we spoke here, as well as all the others
which we came upon on Oahu, were never quite suitable for
temperature measurements. They were either too shallow or
their basins were too large, so that the water in them, which had
been exposed to the sun for a long time, had a much higher
temperature than it should have had.

The valley in which we now found ourselves is quite completely surrounded and it is almost impossible to find an exit
towards the west. Today there is no longer a trace of the many
plantations and homes which are mentioned in King's travelogue.†
Aside from the little village of Mokapu there are no dwellings
here—or at least only very few. This made a continuation of our
journey through this region out of the question since we lacked
provisions and were separated from all our baggage and our instruments. It had been impossible to bring these things down the
steep mountain side. There remained nothing for us to do but
to return by the same way we had come.

The large and magnificent trees of the *Pandanus odoratis-
simus (?)*which could be seen standing here and there were the
chief ornament of this valley in which the higher, tree-like vegetation was quite sparse. The fruits of the *Pandanus* attain the size
of a small head and are frequently used by the natives as ornaments. The lower portion of the cone of this fruit is regularly
square-shaped and all of a golden yellow color. The Indian women
separate it from the green colored portion, thread it lengthwise
and wear the wreaths either around their necks or their shoulders.

We had once again collected a great number of plants and
it became necessary to wrap them up. We sat down in a depression in the shade of magnificent trees of the *Jambosa malaccensis* [42] whose shiny white apples were beginning to redden. Seldom

*Chara oahuensis n. sp. Chara gracilis utriculis simplicibus caule subramoso glabro
ad basin verticillorum ramulorum pilis (Stipulis) adpressis circumdato, ramulis 8-10
articulatis articulis 5-6 apice foliolis 4, 5-6 subfasciculatis sporangium aequantibus
coronatis. Organa sexus utriusque in eodem individuo versantur.*

† James King, author of the last volume of Cook's *Voyages.*

have we wrapped up our collected treasures under more beautiful surroundings.

Late in the afternoon we returned to Madame's Boki's country house in which we had already spent one night. We were more than a little surprised to find a pig baked in the ground awaiting us. Supposedly Kauike'aouli had it prepared for us for our noon meal but his servants demanded more money for it than the thing was worth. The old hospitality of the islanders has disappeared with the introduction of the Christian religion and the desires of the more sophisticated world. During our entire stay on the island of Oahu we did not once, either from the royal family or from the Indians, experience a sign of that old and most praiseworthy characteristic. Kauike'aouli himself never showed us any hospitality — not even by offering us a glass of water.

To our inquiries as to whether from our present location we could cross the mountains which enclosed the valley and thereby continue our journey we received the unhappy reply that this could not be accomplished. We therefore found it necessary to return to Honolulu, where we arrived late on that same evening.

If we finally assemble all our individual observations, we could see that the mountain range which we had crossed on this excursion consisted at its base of a porous basalt which at higher elevations became more and more porous and turned either into real pumice or a pumice conglomerate. Only here and there did small sections of dense basalt or basalt conglomerate emerge. This rock composition can be found on many other so-called volcanic islands of the South Seas, as for instance on the island of St. Helena.

Climbing Mt. Kakea
∽ Five ∽

June 27. In order not to lose any time we set out again today and directed our excursion towards Mount Kakea, one of the highest peaks of the mountain range situated east of the Honolulu Valley.

As we walked through the streets of Honolulu we noticed in almost all open areas and near the houses a great number of *Argemone mexicana*[43] growing like weeds. Among some of these plants stood an Indian woman and we saw her touching the individual blossoms and carrying out some procedure. We approached her and saw to our surprise that she was busy manually spreading the pollen of the blossom onto the stigma. In answer to our question as to why she did this she replied that thereby more seed would be produced. These are eaten as we eat poppy seeds. It would be interesting to know—but not easy to ascertain—whether the inhabitants of the Sandwich Islands had not surmised the sex differentiation in plants earlier than the Europeans.

Later we came by the house of an Englishman in whose yard were seven or eight large land turtles which had come from the Galapagos (Turtle Islands). Upon our request the Englishman sold us the largest of these animals which we safely brought back to Berlin and which is presently in the menagerie of His Majesty the King on the Peacock Island near Potsdam. Since mariners first came to these islands, the Galapagos have become famous for their large turtles. It had been believed that these turtles were sea turtles which came to these islands only at breeding time. This, however, was an error. The turtle which we brought along (pictured in the third part of this book on Table III)[†] is a true *Testudo* and was first described and pictured under *Testudo nigra Quoy* and *Gaim.* [*] This turtle is considered the tastiest in all the South Seas and is valued greatly. The export of these animals from those islands is considerable and everywhere—in America as well as in China—they are offered for sale. In recent times the whalers have begun to stop off at the Galapagos frequently in order to stock up on water, wood and fresh provisions such as turtles and fish. Often their turtle catch is so abundant, that they take them on as cargo in order to transport them to America or to the other South Sea Islands. The animals, which often weigh up to 100 kilos, are then packed on top of each other in the hold of the ship and are not fed during the entire trip.

We had this turtle on board our ship for almost a whole year and often had to marvel at its toughness. It often happened that the animal crept out of its designated place and thereby became bothersome to the sailors. Some of these sailors began to hate it exceedingly and earnestly sought to kill it. First they drove a big nail into its head and on another occasion they drove a large bolt through the hard shell deep into the viscera of the chest. Still the animal did not die and was healed again after eight or nine months. Since the homeland of this turtle lies right below the equator where the mean temperature is at the least 22°R (27°C, 81°F) the animal was very sensitive to the cold. On our return trip in the spring of 1832 it lay for about four months in a kind of hibernation without taking any nourishment. During this time it lost an extraordinary amount of weight; in fact, about 40 pounds. On the Sandwich Islands the turtle weighed 125 pounds, and on our return to Berlin only some 70 pounds. Since

*Freycinet Voyage. Zoolog. p. 174. Tab. 40.

† This refers to Meyen's original German edition.

it has been kept quite warm in the royal menagerie, however, it has gained a very significant amount of weight again but the increase in its size over the last two years is so insignificant that—if one can draw a conclusion from this fact—this animal must be extraordinarily old.

The excursion which we had planned for today, July 27th, took us by the foot of the extinct volcano which lies on the eastern end of the city and is called Puwaina [Puowaina]. This old cone rises to a height of 400 feet and is completely round. On the northwest side the rim has collapsed somewhat. Since the mountain has at present been converted into a fortification, not everyone has access to it but it is not supposed to be difficult to obtain permission. Behind the crater a ridge rises suddenly, forming two valleys which run parallel to the Honolulu Valley in which we had made our first excursion. Soon we reached an altitude from which we could look into the fortification which is situated on the peak of Puowaina. The fortifications consist almost solely of ten or twelve cannons of high but unequal caliber which range over the harbor but cannot be aimed. Every time the current ruler leaves the island of Oahu and again when he returns, he is saluted with these cannons.

The flat valley of Honolulu through which we hiked on this excursion as well as the entire slope of Puowaina and the ridge which we had just climbed were completely barren up to an elevation of 600 to 700 feet—covered only by low herbage* and grasses which at this time of year were almost completely scorched by the sun. Everywhere boulders of porous basalt lay in our path on which we often saw little lizards sunning themselves. These were so extraordinarily quick, however, that we could not catch a single one even though my Indians also pursued them eagerly. Besides these, some large spiders which carried their egg sacs on their backs, some dragonflies and a *Sphinx*,** not much else of note was to be seen. The little children—and often the grown-up Indians too—catch this beautiful butterfly,

*The pretty *Kalstroemia cistoides End. Misc.* (*Tribulus cistoides.*) stood out among these.

**Sphinx convolvuli. (Sph. pungens Eschsch.)

pull his proboscis far out, hold him by it and let him flutter about, whereby the animal makes a constant humming sound. This cruelty to animals is quite similar to that which the children back home indulge in with the cockchafer. [†]

From the mountain ridge we had a magnificent view of the beautiful transversal valley which lay at our feet and ran parallel to the Honolulu Valley. This whole valley is covered with taro plants, bananas and sugar cane and a great many workers were busily occupied there. From this distance the dark brown skin of the naked working Indians and the dark velvet green of the taro fields presented a striking contrast.

On the mountain slope at an elevation of about 800 feet above sea level and just at the spot where there were a few small houses and the arboreous vegetation began we surveyed the vol- canoes which lay to the south. Bejahi [Leahi, Diamond Head] lay south 8° east, Puowaina with the fortification south 40° west and Maunaroa [Moanalua], the third volcano which we will get to know later on, west 10° north.[*] The three volcanoes men- tioned are all extinct and lie almost in a straight line. Only the middle one, Puowaina, lies somewhat more to the south. Mokapu, which forms the northeastern tip of the island, and the fifth volcano, which stands on the northern side of the island, we had already mentioned earlier. These five craters, all of them extinct and probably all inactive since the formation of the island, lie almost in an exact circle. They show clearly how awesome was the power which once thrust the island with its mountain range —whose highest peak rises over 3000 feet—out of the depths of the sea.

Near one of the little huts which stood here we found a lit- tle plant of the *Piper methysticum*. This plant was at one time frequently used in the preparation of the awa drink but its cul- tivation has been pretty much abandoned—fortunately for the Indians.

---

[*]The declination has not been deducted from this data. It was 10 degrees east at that time.

[†]A large European scarablike beetle. *(Melolontha vulgaris)*.

We now walked through the splendid meadows consisting primarily of *Cyperaceae* from whose various shades of green emerged the golden yellow blossoms of a little *Sida* * and the lovely flowers of some *Convolvulaceae*. In tropical regions one must usually do without the sweet pleasures which the sight of the beautiful meadows in the lowlands of our northern regions offer. However, meadows such as these are capable of giving such pleasure. The region of the ferns follows this plain of *Cyperaceae*. Nowhere else have we seen a greater number and variety of these plants, nor have other travellers reported it. The ferns of this region are all arboraceous but they do not have the high, smooth trunks of many of the ferns of America and the Old World which often exceed a height of 20 feet. They are rather more stocky and seldom attain a height of 4 feet. Also the leaf stems never fall off as smoothly from their trunks as from those big trunks; indeed they leave rather uneven remains. Here too we saw the duidui [kukui] tree (*Aleurites triloba*)[45] whose nuts are quite frequently used for burning as lamps. The branches of this tree, which often extend quite far, are connected to the ground by vines** and adorned with thousands of blossoms of the *Convolvulus palmatus*.[47] There are also numerous *Acacia heterophylla*[48] in these forests. Their trunks, which at times attain a diameter of seven to eight feet, usually provide the wood for canoes. Often in these forests one finds whole stretches where there are no tall trees and where everything is covered with these shrub-like ferns whose young, sprouting leaves are enveloped in a long, fine "wool"[†] which the natives gather and sell to foreigners to use instead of hair or feathers to stuff mattresses. An extraordinarily large amount of this fern "wool" is presently being used on the Sandwich Islands. We ourselves have often seen beds stuffed with this "wool" and have tried them out. They are extraordinarily soft, although, we think, somewhat warmer than those of hair. On the other hand the foreigners who have used these beds and mattresses could not praise their coolness

Sida ulmifolia Cav.[44]
**Convolvulus bona nox.*[46]
[†]Called *pulu* by Hawaiians.

enough. This excellent "wool" which has a golden yellow color and is very long is gathered from very diverse ferns. The genera *Asplenium,* * *Aspidium, Davallia,* etc., yield the finest. On the other hand the genera *Sadleria, Acrostichum, Pteris,* among others, yield a much coarser "wool" which is not commonly used.

Finally the foot-path on which we had been hiking up to the top came to an end and we had to clear our way with the greatest effort. Besides the many stems of the *Böhmeria*[49] and *Neraudia* from which the Indians make their tapas, we also saw the *olana*[50] (also *orana*)** tree, whose bast is primarily used for the making of nets and fishing lines and whose large, heart-shaped leaves give it a beautiful appearance. This tree sometimes puts out shoots which grow to 20 feet and more and are completely straight. Nowhere on the island did we find more of those strange *Lobeliaceae* which Mr. Gaudichaud had described, than here. In addition to the already familiar ones we collected a new *Clermontia*** and many other beautiful plants such as *Alyxia olivaeformis Gaudich., Scaevola gaudichaudiana Cham., Vaccinium cereum Forst.,*[51] *Coffea mariniana Cham.,*[52] *Myonima umbellata DC.,*[53] *Charpentiera obovata Gaud.,* a new *Anoda***** and a new genus from the family of the *Rubiaceae,* which we named after our distinguished friend, the younger Mr. Wiegmann.*****

On our way we also saw a little piece of land which was covered with dry taro. It was a damp place. Nearby we came

*Especially *Asplenium patens K.*

**The plant belongs to the *Urticaceae* but we have as yet not been able to determine it.

****Clermontia kakeana n. sp. Cl. foliis late oblongis acuminatis basi attenuatis argute serratis subtus ad venas puberulis.*

*****Anoda ovata n. sp.*[54] *A.fruticosa, foliis obovatis crenatis cauleque tomentosis, pedicellis axillaribus solitariis folio longioribus, fructu mutico.*

******Wiegmannia.*[55] *Rubiacearum nov. gen. Ernodiae proxim. Calycis hemisphaerici limbus quadripartitus, laciniis venosis. Corolla infundibuliformis limbo 4-lobo. Stamina 4, inclusa. Capsula subglobosa, octo costata, laciniis calycinis subfoliaceis sinubus disjunctis superata. Seminibus in loculis solitariis peritropis. Flores involucrati.*

Wiegmannia glauca n. sp. W. suffruticosa caule subanguloso glabro, foliis oppositis, caulinibus subpetiolatis lanceolatis acutis, involucratis sessilibus cordatis acuminatis, calyce foliisque involucratis superioribus glaucis, corolla calyce multo longiore purpureo.

across a spring. They had formed the earth around the root of each plant into a little hollow so that moisture could collect there. The cultivation of dry taro is also native to the Friendly Islands. Forster saw it there.

The top of Mount Kakea, [now known as Sugarloaf], which we reached right after noon time, is bare of all arboraceous vegetation. Bushes six to seven feet in height and connected by an extremely dense growth of *Dracaena* and *Convolvulus* cover the whole area. The last stretch of the way to the summit was so densely covered with plants that we first had to cut a path through them. The dense growth also robbed us of any view. When we arrived at the summit we were hard pressed to gain enough space by chopping down all the vegetation around us to have an unobstructed view and to set up our barometer. The elevation of this mountain, calculated according to Oltmann's formulae is 1596 Prussian feet. Our barometer stood at 28 inches 4 L. English measure at 18.5°R (23.1°C, 73.6°F) mercury temperature and 17.8°R (22.5°C, 72 °F) air temperature.[†]

When the weather is clear the view from this mountain is supposed to be extremely beautiful. From here one can see all the islands over which Kauike'aouli reigns. On this day however, there was a strong sea breeze and the horizon was obscured by clouds so that we could see only the island of Maui.

After we had descended from the top of the mountain we camped in the shadow of the magnificent forests through which we had just wandered. It was high time that we should pack the numerous plants that we had collected, and satisfy our extraordinary appetite. My Indians enjoyed their meal extremely well and we were amazed at the quantity which they were capable of consuming at one sitting. The reason for this may of course be the minimal nutritive value of the taro root, their customary food. As far as we know, even our potato is more nutritious than the taro root—though not as tasty. By noon the people had consumed enormous calabashes of poi, that beloved paste. Indeed

[†]The Reaumur scale which is still used in Europe was invented by Rene Antoine Ferchault de Reaumur in 1731. The freezing point is 0° and boiling is 80°.

they were never satisfied until all the food intended for the day had been eaten. The meals were always spiced with witty phrases and funny faces which the whole group applauded with laughter.

After we had left this resting place we sought to return to Honolulu by a different way. We climbed down into the valley which ran along the left side of the ridge on which we had climbed up to the summit of Mount Kakea. The mountain had an incline of 65° and an elevation of more than 500 feet but we were nevertheless able, with the help of the extremely luxuriant vegetation, to descend on this steep slope. Our endeavor was most difficult and in many places almost impossible; but to see this indescribably great quantity of herbaceous, tropical vegetation at close range was sufficient reward for our venturesome undertaking. Several of my people tried to find a better way at other points but they were completely cut off from us when the dense forests and tall plants which surrounded us everywhere removed them from our view. They had to turn around again later on, and they returned to Honolulu by a completely different way.

Nowhere again, neither on Oahu nor in Brazil nor in Manila, did we see such a charming picture of nature. We saw here the greatest profusion of the gayest tropical vegetation complemented by the picturesque forms of the mountains. Numerous *Musaceae*, some casually planted, others wild, covered the slope of the mountain. Among them were the fragrant and aromatic *Scitamineae* which were already mentioned above (p.21), and also the short, shrub-like ferns intertwined and covered with vines which had blossoms of the most wonderful colors. Beneath that were the various greens of the *Cyperaceae*, which cover the lowest parts of the transversal valley, as well as the loveliest arrangement of the individual clusters of shrub-like and arboraceous vegetation on the slope of the mountain ridge and on the top of the mountain close by. All this taken together made such a glorious and friendly impression that we were often not capable of going on. Had it only been possible to have a view of this region — even if only a small portion of it — copied by a talented artist!

It is striking how nature, in bringing forth certain forms of her plant and animal kingdoms, is so exactly bound to localities, for reasons which we have not begun to suspect. The forests of Brazil abound with ugly amphibia and countless insects. Seldom does one touch the branch of a tree or the leaf of a plant without coming upon various sorts of insects. But here on the island of Oahu, as on the rest of the South Sea islands, there are very few insects. In vain one examines the underside of leaves, in vain one shakes the trees—no insects fall down. One does, however, come upon snails in pretty shapes and often in the most brilliant colors. Often they have regular stripes and remind one of our *Helix nemoralis*. Some are completely grass-green but this color disappears at death and has probably only been transferred to the shell by the eating of green leaves. Here on the Sandwich Islands nature has placed countless land snails instead of insects on the leaves of the trees. In the East Indies she observes the middle ground. There, as for example in Manila, she has assigned partly land snails and partly insects to the vegetation—both frequently of enormous size and the most brilliant colors. The diversity in respect to size, color and shape among the land snails of the Sandwich Islands is extraordinarily great. Mr. von Chamisso* has already described an *Auricula owaihiensis* and an *Auricula sinistrorsa* and Mr. Green** an *Achatina Stewartii* and an *Achatina Oahuensis,* besides which several new types were brought back by French naturalists and by ourselves. The number of these snails is exceptionally large and they vary in respect to size and coloration quite extraordinarily. It seems particularly noteworthy, however, that the greatest number of these snails is coiled to the left while at home and in all other regions this abnormality is very rare. Indeed there are species of the genus *Achatina* which seem to occur only coiled to the left on the island of Oahu.

Oftentimes we felled great stems of the *Musaceae* in order to examine the sheaths of their leaves for insects, but aside from a few earwigs and a *Blatta,* which was probably *Blatta orientalis,* we saw none.

Nova Acta. Acad. Caes. Leop. Tom. XIV. P. 639.

**Stewart's *Journal of a Residence in the Sandwich Islands.* In the appendix. J. Green, A. M., Professor of chemistry at Jefferson Medical College, Philadelphia.

As we descended farther down into the charming valley the small stream which flows in it became larger and larger. Some Indians had built their huts beside it and had prepared some land for the cultivation of taro. The footpath on which we wandered through the valley was extremely difficult for us to negotiate. It often ran along the slope of smaller and larger rises which were covered with very smooth *Cyperaceae* and offered no firm footing. When someone from our company fell, or even just slipped —which happened quite frequently—the whole group laughed quite heartily. Our company always made a short stop then to exchange a few jokes. Happy that they were going home again, the Indians sang their monotonous songs which sounded as though the pupils in our elementary grades were singing the ABC. Now and again one of the Indians would display his greater singing talent either with a trill or by drawing a tone on into the next verse, which was acknowledged with general laughter.

As soon as the valley became wider the beautiful vegetation disappeared. The slopes of the mountains were covered only with low grasses, the huts of the Indians became more numerous and here and there large boulders appeared again. The end of a low ridge which runs through the center of this transversal valley had been artificially cleared of vegetation and of the cover of humus. The rock which came to light here is a very attractively colored basalt conglomerate.* The Indians were just then busy chipping flat pieces from this rock which they wanted to use to hunt octopus. The rock on the sides of the valley, however, is the usual porous basalt which is found all around Honolulu. Here and there one can find caves in this rock, some of which are inhabited.

In the course of our excursion we saw the mountains everywhere covered with grazing horses and horned cattle. One is amazed at the great number of cows which thrive here beautifully with the slightest care. It is moreover an excellent breed with

---

*Basalt conglomerate: small angular pieces of the basalt mentioned on p. 34 which lie quite close together are connected with each other through white calcite. The calcite is fine grained but often forms little hollows whose walls contain little crystals which seem to be the second-order six-sided prism, pointed with the surfaces of the first more pointed *rhombohedron*.

beautiful long horns which was imported here. The island of Oahu has more than 2000 head of horned cattle of which 1000 head belong to the Spaniard Don Francisco Marin. On the island of Hawaii, where Vancouver is known to have imported horned cattle, they have grown quite wild and love to roam in the colder regions of the high volcanoes of this island. They live there in great herds and the inhabitants actually hunt the animals. We were told that these great herds at times come down from the mountains and into the villages of the Indians who must then flee. There is also a great number of horses on these islands and already every reasonably well-to-do person, man or woman, keeps a riding horse. Yet, as welcome as the increase in this most useful domestic animal is, the joy in it will soon disappear when it is realized that this increase, as well as the expanded cultivation of meadows, is in exact proportion to the decrease in true agriculture.

Everywhere one hears the complaint that in former times a far greater quantity of field-produce was cultivated than now. This complaint is justified, even if the missionaries try every possible means to free themselves in their writings of this weighty charge. Many and very extensive fields through which we have just wandered and which are presently being used as pasture land were formerly covered with sweet potatoes.[56] Today one can still see the remaining traces of their cultivation. They say that in the days of Kamehameha a great part of the Honolulu Valley was used for the cultivation of field-produce. Now there are meadows there and the valley is far less productive than in former times.

Finally we reached the plain of Honolulu and looked out across the sea where several ships flying the flag of the Sandwich Islands lay at anchor beside our *Prinzess Louise*. Here stand several estates and summer homes of wealthy foreign merchants and wonderful fruits from their fields were being offered for sale. The melons were extraordinarily large—almost as large as our large pumpkins—and yet their smell and taste was most pleasant. Here as in Chile they are so sweet that one can eat them without any sugar.

At sundown we returned to Honolulu. Our way took us by the residence of King Kauike'aouli. He, along with some friends and court favorites, was playing a ball game which is very popular on Oahu.

A Royal Banquet
~~ Six ~~

June 28th. In accordance with the orders which had been given to Captain Wendt, Kauike'aouli was invited to dinner on board the *Prinzess Louise* today. He appeared in the company of his uncle, Governor Kuakini, Kaiki'oewa, the governor of Hawaii, his secretary Halilei and some court favorites. General Miller and several other foreigners had also been invited. Kauike-'aouli came on one of the boats of the *Prinzess* and brought several servants along who carried poi and taro, the customary foods of the Sandwich Islanders, so that if he still had an appetite for these after the meal he could satisfy it immediately. The young ruler appeared in the same dress which he had worn at the presentation of the gifts. His servants, however, carried the uniform which had been among the gifts we had presented to him from His Majesty, the King of Prussia. Shortly before dinner, Kauike-'aouli changed into his uniform but we noticed that the sword, the plume on the hat and the spurs were missing. He told us that his servants had forgotten them, without, however, rebuking them for it. This was a concocted excuse though, for we soon learned from Halilei, his secretary, that Kauike'aouli had purposely left these things at home, as the missionaries had told him that wearing these things would be an insane and most indecent thing to do.

The banquet on board the *Prinzess* was held on deck under a canopy. The guests conducted themselves quite well at table, although they ate a shockingly great amount. Often, though their plates were still full, when they noticed that any bowl was about to be emptied, they quickly had themselves served from it. In their drink they were quite moderate, but then their capacities were quite great. Kauike'aouli, who drank only Madeira wine, made a toast to the health of His Majesty our King by exclaiming, "The King of Prussia!" and, with the thunder of the gun salute, there was never a more joyous toast among Prussians. Soon thereafter Captain Wendt also honored the young ruler of the Sandwich Islands with a gun salute.

During the meal the guests brought up various complaints about the present government of the Sandwich Islands. Kauike'aouli, who is quite familiar with these but does not have the power to bring about any changes, turned away from this conversation, which he quite liked to hear, and let all the abusive language fall on Governor Kuakini, the brother of Ka'ahumanu. Secretly, he said, however, that everything would improve when Boki returned. Kauike'aouli harbors this hope in vain, however, for Boki will never return. It is now certain that he was shipwrecked.

Boki was a brother of the deceased minister Kalaimoku. He and his wife accompanied King Liholiho to London and after the death of his brother became Governor of Oahu. Boki protected the young King against the power of Ka'ahumanu and the eries [aliis], as well as holding the missionaries in control. About fourteen months ago this man was sent to the New Hebrides on the brig *Kamehameha* which the king had bought for 40,000 Spanish piasters and had paid for with sandalwood. Allegedly Boki went to get sandalwood but he had 360 armed Indians on board and so he probably went only for conquest. The brig *Kamehameha* was lost, as we learned on our return to Europe, and the loss which the Sandwich Islands thereby suffered is very great. The result of this abortive undertaking was that Kuakini, formerly Governor of Hawaii, came to Oahu bringing his own soldiers and is now, with the help of his sister, the old Queen Ka'ahumanu, the actual ruler of the Sandwich Islands.

After the dinner Kauike'aouli changed his clothes again and left the *Prinzess Louise*, which gave him a seven-gun farewell.

Waikiki and Diamond Head
≈ Seven ≈

June 29th. Today we visited the eastern side of Oahu on horseback in the company of General Miller, Dr. Ruck [Rooke] and several Indians. Our way took us through the plain along the beach which was only sparsely covered with grass. Not until we came to the village of Waikiki, where there is flowing and standing water in abundance, did we see the taro fields and precious coconut plantations which stretch almost right up to the ocean shore. Under the scant shadow of these trees stand the quaint huts of the Indians. We had some coconuts picked for us in order to quench our thirst with their famous milk. The Indians scrambled up the extraordinarily high trees quite like monkeys. They placed the forepart of the foot on the rings of the palm trees and, holding the sides of the tree with their hands, they went up much faster than we could with our method of climbing. Here one must note, however, that the trunks of the coconut trees are particularly suited for this and also that they do not let them grow straight up but try to pull them somewhat to the side, which facilitates the climbing. Because the Indians always go barefoot and also practice this kind of climbing, their toes become so flexible that they can be used similarly to the fingers. As soon as the Indians had brought the nuts down, they took the nearest available stone and broke open the soft hull of the nut, which was then pulled off with the help of their teeth. Next a stone was driven into the soft upper part of the nut itself and the milk drunk out of it. To us the cocount milk, which is so often praised, seemed to be quite a tasteless drink. It is similar to our sweet whey.

Not far from the coconut plantation there were small pools of water which were densely filled with a *Scirpus** to which Mr. Nees von Esenbeck has given our name. It is this *Scirpus* from which the fine blankets are made that are so rare at home and even on the South Sea Islands so extraordinarily expensive. Mr. Nees von Esenbeck took seeds from this plant in our herbarium and planted them and now this interesting and useful plant is being cultivated in the botanical garden in Breslau. Already in the first part of this book[†] we mentioned a similar plant which covers the shores of Lake Titicaca. Without it the people of that region would be quite unhappy. Here on the Sandwich Islands these rushes serve as a means to the good life. Only mats of secondary quality are made from this *Scirpus*. Coarser ones are made from other *Cyperaceae*, the coarsest from pandanus leaves and the finest from another, very fine *Cyperaceae*** species. To prepare the fine mats the grass stems are first carefully dried. Then the hard outer layer is separated in narrow band-like strips from the cellular tissue which fills the inside of these stems. Each strip is then carefully smoothed and scraped until it becomes a gleaming white. The work is so infinitely tedious, that a woman who is practised in it is said to be able to complete no more than one mat eighteen feet long and eighteen feet wide in the course of one year. Such a mat is then sold for 14 to 15 piasters. Smaller mats of the same type, eight feet square, are worth 5 to 7 piasters. Very seldom do they make multi-colored mats on Oahu, whereas it is precisely these which are in common use on the Friendly Islands [Tonga]. Indeed we feel that there they have advanced

*Scirpus meyenii Nees v. Es.[57]
**Eleocharis palustris n. Sp.[58]
†This refers to page 491 in the original German edition which includes his travels in South America.

somewhat further with this craft—at least the prices for equal quality articles are much higher on the Sandwich Islands than on the Friendly Islands. The use of braided mats is extraordinarily great on all the South Sea Islands. They are used to cover the ground on the inside of the huts—even in the home of the poorest Indian one will not look in vain for this furnishing. Furthermore these mats are used as hangings that divide the inside of the houses into different rooms—indeed they even sometimes form the outer walls of the homes, which provides cool air on the inside. Mostly these braided mats are used for sleeping. For this from 7 to 12 mats and often more are laid one above the other, the coarsest on the bottom, the finest on top. One then lies down on these without further preparation, usually in one's daytime clothes and without a cover. Only the more refined have pillows and these are made of the fine fibers of some ferns of which we spoke earlier.

At the edge of the village of Waikiki we noticed several quaint huts which stood somewhat apart from the rest. This is a place of exile for Honolulu.

Our way took us right along the beach which, like almost the entire Honolulu Valley, is covered with coral rock. We noticed that a number of Indians, both men and women, were playing in the surf. They approached the surf very slowly, swimming on a board, and by waiting for the right moment and through a skillful movement they were able to suddenly lift themselves over it, without being hurt by the falling mass of water. The Indians swam against the surf at the same moment that the wave rose up and with lightning speed they shot through the tall wave before it fell again. Other Indians sat in their canoes and were entertained by this and still others were busy right next to this activity with fishing or were competing with each other in fast paddling.

With this entertainment we reached the foot of Diamond Head and were most pleased to find the old morai [heiau] there still quite intact. Until the introduction of the Christian religion, all the sacrifices and ceremonies of the former worship on Oahu were practiced on this morai, which is an area more than 200 feet long, 100 feet wide and surrounded by a wall of basalt rocks 7 feet high. The side which faces the ocean is terraced but is also enclosed by the wall. Within the enclosure we saw small piles of stones lying here and there but bones were no longer to be found, although it is certain beyond doubt that before their encounter with the Europeans these people, too, made terrible human sacrifices to their gods. The high priest from those times is still alive; in fact, he lives on Oahu. He talks freely about the ceremonies of his former religion. Not without shuddering we left this place —this evidence of strange aberrations of the human spirit—and rode further east, always close to the beach which was quite empty of the things usually tossed ashore by the sea. Frequently we came upon the shells of the large and beautiful *Palinurus penicillatus Oliv.* which is quite numerous on the coral reefs. Its size makes it a valuable food. This crab varies extraordinarily in its coloration. Now it is blue, now reddish, now greenish or many other colors, so that one does not find two specimens that are exactly alike. It is also extremely rare, here as in other regions, to find these large *Palinuridae* intact. Sometimes their antennae are broken off; sometimes they are missing several feet. Besides some shellfish like *Cypraea mauritiana, C. tigris, C. arabica, C. guttata, C. isabella, Bulla amplaustra, Buccinum maculatum,* several *Murex, Conus, Nerita, Patella* and *Turbo* types, as well as some very small seaweeds, there was not much else to see. A beautiful *Convolvulus** grows down into the plain here and close up to the beach. With its creeping stem it enlivens the sandy region. On the southern side of Diamond Head, right by the sea, some basalt lies exposed which trends in layers to the southwest and disappears under the trass which alone forms the hills of this region. We climbed up Diamond Head, an extinct volcano, whose crater rim is somewhat collapsed on the southeastern side. The mountain is very likely over 500 feet high; its crater is about 100 feet deep and had, when we were there, a small pool of water at the bottom, which was completely covered with plants. The mountain consists of brown basalt tuff** which, as Hofmann

C. ovalifolius V. [59]
**Basalt tuff. A sample piece had on the one side the appearance of the basalt

had already noted earlier, contains veins and pockets of lime. This trass is here and there more or less distinctly layered and the layers are of varying thicknesses. The diameter of this crater, the rim of which is very narrow, we estimate to be about 1000 feet across. Here on this old volcano we collected extremely beautiful plants among which a *Gouania** and an *Euphorbia*** stand out especially. East of Diamond Head the land becomes flat again. The high mountain range from the center of Oahu ends at Diamond Head and becomes very flat, with only a ridge running out to the south point of the island. There stands Koko Head, also an extinct volcano, which we mentioned earlier.

Somewhat further to the east, close to the ocean shore, we found a spring. It had only a little water but it was mineral water. Numerous small crabs ran across the sand with the greatest speed, either seeking their holes or attempting to escape pursuit in the ocean. It is most delightful to watch these animals running on the tips of their long legs and moving sideways. Now and again we came upon individual Indian huts with some *Ricinus* stems, or cotton plants, standing next to them. Whenever we stopped at any of these huts, we were offered pierced shells, pieces of tapa (pieces of cloth made of tree bark), and other things of no value for sale. Though one could enter everywhere and sit down, there remains no trace of the former hospitality for which these islanders were so well known. They sold us bananas and watermelons, but then ate even more of these than if they had still been their own property. We in no way want to attribute this change in the kanakas to a degeneration in their nature, but rather

conglomerate mentioned on p. 46. The enclosed basalt pieces surrounded many olivine crystals, which also occur alone in the calcite. They are small but usually nicely and clearly crystallized. Their form is that which olivine usually takes in basalt. On the other side of this piece the mixture becomes finer and takes on a fairly regular brown color. The basalt pieces have disintegrated and the whole thing is bubbling with acids.

*Gouania integrifolia n. sp. [60] *G. fruticosa, erecta, simplex, foliis longe petiolatis coriaceis subrotundis retusis integerrimis subtus petiolisque puberulis, axillaribus paucifloris petiolo brevioribus, calyce hirsuto.*

**Euphorbia cordata n. sp. [61] *E. Fruticosa, humilis, ramis divaricatis, torosis pubescentibus, foliis oppositis confertis subsessilibus cordato-ovatis obtusiusculis mucronatis integerrimis glabris, coriaceis, involucellis terminalibus solitariis campanulatis ore villosis.*

solely to the oppressive poverty which these Indians presently suffer. The foreigners not only introduced them to the luxury articles of the civilized world, but have also commanded them to wear clothing which they cannot afford.

When the noon heat became too great, we rode back to Honolulu. Everywhere in the small streams and pools which we had to cross we startled Indian women, who were enjoying the cooling water.

While we were still on our way, some strange news was brought to us, which agitated almost all the Indians on the island. The news was that Boki, the former Governor of Oahu of whom we spoke in some detail above, had suddenly returned on the brig *Kamehameha*. A poor Indian from Waimea Bay brought this news to Honolulu and added, that he had seen Boki himself. Since the news was extraordinarily important to all who inhabit the Sandwich Islands, messenger upon messenger was immediately dispatched to the northwest side of the island in order to confirm it. Madame Boki, the wife of the expected Boki, immediately mounted her horse and rode out to meet her husband. Towards evening the dispatched messengers returned and declared the whole report of Boki's return to be fictitious. The poor Indian who had spread the news was immediately seized and thrown into the fort only to be dragged out a few days later in a terrible manner and his staring eyes were sufficient proof for us that he was mentally ill and therefore not accountable. Besides, he had no personal interest which would cause him to spread this story about Boki's return. But the man who had conjured up before the sinners the punishing spirit had to be punished by them.

One morning the punishment for that alleged lie was inflicted upon this feeble-minded man in the streets of Honolulu. With his arms and chest bound to the back of a cart he was dragged along. On the cart sat an official with a switch in his hand and everywhere the cart—which was pulled by Indians—stopped, he informed the people of the captive's crime. Hereupon they always bound the captive's feet to the wheels of the cart and flogged him anew. When we saw this man whose back, covered

with blood, was receiving fresh blows, we had to turn our eyes
from this scene of misery. An old Caucasian with white hair and
a snow-white beard—an Englishman by birth—was the executor!!*
Thousands of Indians, young and old, men and women, followed
this scene of misery and, like spoiled children, exclaimed their
joy when the sick man screamed terribly. Alongside this procession walked a number of the naked soldiers of the Governor.
Most of them had a wife beside them, a musket in one hand
and a naked child on the other arm. Among such a good-natured
people as the kanakas, who are also at a low level of civilization,
one constantly finds opposite extremes side by side. In the past
they let themselves be sacrificed to the gods by their priests and
now they are allowing themselves to be beaten to death for an
alleged lie. May the lies which the missionaries on the Sandwich
Islands have deliberately sent out into the world be punished less
harshly but may those who utter them unknowingly be completely forgiven. May the lies be charged only to those people
who sent out the missionaries and were not able to choose better!

Quite late that evening Captain Wendt and I visited Governor Adams, who makes his home in the fort of Honolulu. We
found him sitting in a chair in the open court yard, surrounded
by more than a hundred of his servants and soldiers, who had to
entertain the Governor with loud conversation and witty sayings.
It was a splendid evening. The moon shown so brightly and the
air was so mild, that we often envied the inhabitants of these
islands for being able to live here.

This kind of evening entertainment, which John Adams was
just then enjoying, is customary among the well-to-do of the
Sandwich Islands. Soon after dinner the people gather around
their patron, sitting around him in a circle and trying to shorten
the long evenings through conversation. Song and dance, like all
lively expression of joy, have disappeared from the huts of these

*Note: On the last day, a few hours before the *Prinzess* put out to sea again, we
had the opportunity to see this old Englishman on his deathbed. He had succumbed
to an inflammation of both lungs and had sent to us for help. Here, in this sad state,
from which he probably did not recover, he admitted that he and another Englishman, whom he named, were the greatest culprits who lived on the Sandwich Islands.

people since the converters of the heathen, through the weakness of an old queen, have been in power on these islands. We had not been there long when we noticed a large, ill-shapen mass lying on a mat on the ground, which was slowly raising itself up. It was John Adam's wife, if we are not mistaken; a veritable monster in circumference and ugliness and also descended from the royal family. She did not speak; indeed she could barely stammer and throw back the blanket in which she was wrapped in order to extend her hand to us.

To Pearl River
～ Eight ～

June 30th. We used this day to make an excursion to the
Pearl River, which runs through a valley and into the ocean at
the southwestern corner of the island, about three miles from
Honolulu. One can go there either by sea in a boat, in which case
one then travels up the length of the river, or, if we may express
it this way, half by water and half by land on a horse. We took
the latter way in the company of the charming Dr. Rooke.[†] This
way took us for more than an English mile through the royal fish
ponds which at low tide are easy to cross and thus shorten the
way considerably, though the horses were up to their bellies in
water the whole time. These fish ponds are large water basins
which are situated right next to the ocean shore but surrounded
by walls of coral rock and thus separated from the sea. Various
small holes at the bottom of the wall allow the fish to freely pass
from the sea into the basins but they are then prevented from
returning. In Lord Byron's[*] account one can find an illustration

* Byron, Lord George Anson. *Voyage to the Sandwich Islands in the years 1824-
1825.* London 1826.
†Dr. Thomas Byde Rooke, 1806-1856.

of these royal fish ponds. At high tide one must make one's way further inland, where there is not as much water but there are deeper ditches and even small and very deep streams to cross.

As soon as one has passed through this unpleasant waterway one enters fruitful and well cultivated valleys along a stream which empties into the sea. We stopped off at a hut in which two canoes were being outfitted with provisions for a long journey. The woman of this humble hut lay stretched out on a mat in the middle of the hut. Another woman who was just preparing to leave went to her and bent over her to press their noses together in farewell. All this happened without a word being said. The departing woman got up and lit her pipe, whereupon the hostess began to wail and scream violently, without shedding a single tear.

For a while we rode along the stream which was lined by beautiful vegetation. Several Indians came by and offered pearls for sale. They asked one real (6 Silbergroschen) for 4 or 5 pearls but they were small and of inferior quality. After a half hour we entered a wide valley which was covered with an extraordinary number of food plants. Bountiful taro fields covered the plain and countless coconut palms,[62] with several huts in their shade beautified the country side. We stopped off at the home of some friendly Indians and quenched our thirst with a watermelon— which we always preferred to coconut milk. In the huts of this fertile area we also saw some pigs and little dogs, which were being fattened. The former are very clean little animals here on the Sandwich Islands and one frequently sees women holding them in their lap or in their arms and letting them eat from their hand. At least such friendliness is extended here just as frequently to the little pigs as to the dogs.

From these last huts our path went steeply up Mauna-roa [Moanalua], an old extinct volcano with an elevation of about 300 feet. The eastern slope of the volcano consists solely of lay-ered trass of a grayish-brown color* which trends northwest. Other places, especially the inner and upper rim of the volcano, consist of the bluish-gray blistered basalt** of which all the

---

*Brown basalt tuff with a brown earthy, slightly ferriferous surface coating.

**Basalt, gray, somewhat porous with many extremely small, white particles mixed in, which seem to have a feldspar-like component.

mountains of the island are formed. On the top there was a crater-shaped depression containing an accumulation of salt water, which is supposed to be covered by a crust of salt during the dry season. On the northwestern side lie two mountains which are the remains of the rim of an old crater. The one further to the north is situated higher and consists of thin layers of blistered basalt which trend about 18° northeast. The layers of the other mountain which is situated more to the south are completely parallel and run horizontally. These two peaks are given the name Moanalua, which means "two mountains." We think that the old crater of this volcano lay to the west of these two peaks and that therefore the basin of the lake is by no means the former crater. Mr. Hofmann* had found olivine and stilbite in the basalt of this mountain. The former, as well as augite, is among our specimens. The route which goes over Moanalua is very monotonous and covered everywhere with boulders. Further on, where the rock is weathered, the vegetation is more abundant. Indeed, the rock of Moanalua is the least weathered of the four volcanoes which we saw on Oahu. Perhaps it was also the last one to erupt.

From this mountain on we progressed quickly, although there was not even a suggestion of a cleared path. We saw several plantations of *Böhmeria albida Hook.*,[63] that plant which is usually used to make the fine tapas. This plantation had a very neglected appearance. Many young saplings could not be seen for the weeds. They do not let these little trees become very tall; rather they prefer to use the young saplings when they are 2 or 3 years old. Presently we reached the valley in which the Pearl River flows. At its mouth this river is more than three leagues wide and it runs from south-southeast to north-northwest. For the first two leagues near the coast the whole valley is very low but then it gradually rises more and more and runs through to the northwestern tip of the island. On the western side of the island arises a separate mountain range which is covered with the most magnificent green. This range does not attain the elevation of the eastern range which runs the whole length of the island.

*Karsten, C. J. B. *Archiv F. Bergbau,* Vol I, issue 2, 1818, p. 301.

Our short stay on Oahu did not permit us to visit this western mountain range. It has not, we believe, been visited by naturalists and is the only place on the island of Oahu where one can still find some sandalwood. At the mouth of the Pearl River the ground has such a slight elevation, that at high tide the ocean encroaches far into the river, helping to form small lakes which are so deep, that the long boats from the ocean can penetrate far upstream. All around these water basins the land is extraordinarily low but also exceedingly fertile and nowhere else on the whole island of Oahu are such large and continuous stretches of land cultivated. The taro fields, the banana plantations, the plantations of sugar cane are immeasurable. Near the homes which form the village of Mannoneo [Manoneo] stand coconut palms, shaded by the magnificent foliage of the breadfruit tree. Here is also the wealthy estate of Don Francisco de Paulo Marin, a man of ordinary education but of noble sentiments, whose name will always rank foremost in the history of civilization of the Sandwich Islands, even when the names of the missionaries will have long been forgotten. Marin introduced the most useful plants from around the world to the Sandwich Islands. The cultivation of these plants will one day be a source of great wealth for the Sandwich Islanders. The Guatemalan cocoa which Marin cultivates is one of the highest quality and perhaps equal to that from Manila, which, because of its high price, does not even appear in our commerce. The coffee tree, limes, oranges, fine grapes, a beautiful papaya which was brought from the Marquesas, the tamarind, cotton, the most beautiful pineapple and many other fruits can be found on the estate of this Spaniard who was once a counselor to King Kamehameha I. Indigo was brought from Batavia by Mr. Serriere and grows to extraordinary perfection on Oahu; however, it is not allowed to be cultivated on a large scale. Neither is sugar or coffee, though it would give thousands of idle Indians employment and good nutritious food. A sugar mill which was in operation here earlier had to be shut down again. Only the ignorance of the missionaries, only their lack of a general education and knowledge of human nature, could have brought about such absurd management.

The estates of Marin are extremely well managed and could serve as a model establishment for the whole land. Several of the enclosures are made of the *Cactus ficus indica*[64] which, all covered with flowers, gives an extraordinarily beautiful effect.

We must mention here another matter which is very strange to the North European when he visits tropical regions. Alexander von Humbolt had already noted that the inhabitants of Mexico, Peru and New Granada, just like the Spaniards in their homeland, do not enjoy the comfort of shade-giving trees near their homes. This dislike for shade trees near their homes is not only characteristic of people of Spanish descent, however. It is almost as common on the Sandwich Islands. Sometimes the Indians on Oahu plant some trees near their huts but then these are the kind which do not give shade, for example, the *Ricinus*[65] trees, cotton, and the *Cactus ficus indica.*[66] Meanwhile the greatest abundance of shade-giving trees is available in the nearby mountains, even at an elevation of only 200 to 300 feet. In very humid tropical regions, where there is a great number of insects, it would of course be very wrong to plant big trees near the house. Not only do they draw insects, snakes, and other ugly animals, but it reduces the breeze and the house soon becomes dank. On the Sandwich Islands these reasons for not planting shade trees do not exist. The lack of insects on these islands is quite amazing. Not even mosquitoes, the common pest in tropical regions, are native to the Sandwich Islands. They have been introduced through the traffic from America but have spread very little so that one can still sleep here quite undisturbed. According to Mr. Bligh there were no mosquitoes on Tahiti either before the arrival of the Europeans who introduced them. *

*Note: Equally conspicuous is the small number of birds to be seen on the Sandwich Islands. We saw only three land birds on Oahu, namely the *Nectarina flava,* the *Fulica chloropus,* and a white owl. In Byron's Voyage, however, there is a very complete list of birds which was prepared by Mr. Bloxham, who added many new species. Mr. Bloxham, who accompanied the *Blonde* as the ship's chaplain, observed the following birds there: *Nectarina nigra, N. Byronensis, N. coccinea, N. flava, Gracula longirostris, Muscicapa sandwichensis* L., *Loxis psittacea* L., *Fringilla rufa, F. sandwichensis, Turdus sandwichensis, Corvus tropicus* L., *Fulica chloropus* L., *F. atra, Solopax solitaris, Procellaria alba* L., *Tringa oahuensis, Sterna stolida, St. oahuensis* and wild geese and ducks.

We had to roam about for a long time before we found a house where we could stop off. In the home of an Englishman, who had come to the island as a sailor and settled there, we found hospitality. The poor natives can no longer offer it, as they are too poor and have almost nothing on hand. Right after our arrival a pig was killed, singed at the fire, and expertly dressed with the help of sharp basalt stones. Then a pit in the ground was heated by means of hot stones, the pig was wrapped in banana leaves and after its cavity was filled with leaves of the taro plant *(Arum macrorrhizon)* [67] —which are enjoyed as a green—it was placed on the stones. The space around the pig was filled with taro roots and then the whole thing was covered and earth was thrown over it. With this type of roasting, which was already described in detail in Cook's voyages, the foods become extraordinarily tasty. Good water in coconut shells and poi served in large calabashes increases the number of dishes at this interesting noon meal, which was served on the ground on large banana leaves.

We could not stop admiring the great fruitfulness and the wealth of edible vegetation in this region and greatly regretted that the short time which we had did not allow us to spend more time there. Sadly we looked once more towards the forests on the western mountain range, which have probably not been explored by any botanist, and quickly set out on our return trip in order to reach Honolulu before high tide. Already we had to ride through several water basins where the water came up to the saddle. Even in the city of Honolulu, at its western end, there is such a basin. It is about 30 paces wide and the natives must constantly wade through it. To do so they take off their few clothes so that these will not get wet.

On that same evening Captain Wendt and I visited Kauike-'aouli, in order to receive the letter of reply which he wanted to send along to His Majesty our King. Kauike'aouli lay in his big house on a bench and two or three of his friends were enjoying a meal with him while his servants lay around him on mats and entertained him with stories. Kauike'aouli said that he had written the letter but didn't know where it was. Not until two days later did Captain Wendt receive the letter and could only then leave the island.

Subsequent Observations
⁓ Nine ⁓

Subsequent Observations about the Situation on the
Sandwich Islands, Particularly in Reference to the
Island of Oahu.

As much as has already been written and repeated over and
over again about the Sandwich Islands, we nevertheless, for various reasons, consider it our responsibility to expand the literature about these islands with several observations and reflections,
in order, on the one hand, to refute at least the major miscon-
ceptions which have crept into some of the more recent geograph-
ical works, and on the other hand to confirm the ill opinion which
the majority of our good readers may have already formed from
earlier travel accounts of the Sandwich Islands about the dread-
ful affairs of the missionaries there.

From Cook's time on, the population of the Sandwich Islands has always been exaggerated. Indeed it has been exaggerated in such a manner that in recent times they have had to invent
all kinds of hypotheses, in order to be able to explain the alleged
great depopulation, which these islands have supposedly suffered. No one, however, has attempted to prove this even to the
slightest probability and we consider it completely unfounded.
In the year 1825 Mr. Hassel* allotted the Sandwich Islands a

*Geography of Australia. Weimar 1825. p. 831.

population of 399,000 which he found, he believed, to be quite appropriate for the area of the Sandwich Islands. Soon afterwards there appeared, through Mr. Stewart,* the first information about the population of these islands which to a certain extent approached accuracy and was here and there based on an actual count of families. However these figures which cite a population of 140,000 for the entire group of islands, are probably also much too high. For example, Mr. Stewart cites a population of 20,000 for the island of Oahu, which is 46 English miles long and 23 English miles wide, whereas in the year 1831, by which time fairly accurate figures had been obtained, especially through the levying of direct taxes, it had a population of only 17,000 to 18,000. If we reduce the population of all the islands by a similar amount, then we would arrive at a population of 130,000 at most for the entire group of islands for the year 1831.

The Indians of the Sandwich Islands, who call themselves kanakas, belong to the major race to which Mr. Bory de St. Vincent happily gave the name Oceaners and differ significantly from the Malayans. It is one and the same race which inhabits the Sandwich Islands, the Marianas, the Carolinas, and the Friendly Islands and to which the Tagalogs of the Philippines also belong, as we have attempted to show in greater detail in another part of this work. **

The family of the alii which is in many respects extraordinarily peculiar, deviates considerably from the common kanakas. All travel descriptions speak of the women of the Sandwich Islands, who are depicted as being so extraordinarily huge and corpulent. These observations are by no means correct, however, for they apply only to the women of the alii, that royal family, whose giant stature in height as well as in corpulence is also characteristic among the men. In all our excursions on the island of Oahu we never saw a kanaka, man or woman, who would have been conspicuous in their corpulence. The extreme corpulence

*Stewart, C. S. *Journal of a Residence in the Sandwich Islands,* London 1828.

**See the third part of our account, page 171: Concerning the Tagalogs in the Philippines as racially related to the Oceaners. [Meyen's original German edition].

among the women of that royal family does not consist of an accumulation of fat, as is the case in many other nations, often to an extreme degree. Rather, their height, as well as their corpulence is based on their bone structure and thus inherited. These people always measure in height 6 feet, 2 inches to 6 feet, 5 inches—sometimes much more—with a corresponding circumference, which is generally not the case among the giants in our lands. These women, with their more or less copper-colored complexions and their formidable figures, often possess great charm—especially because of their very lively, shining black eyes. They have hands and feet which exceed in size those of the biggest German sailors. The feet of these people in the region of the ankle are so extraordinarily large, that they appear quite unnatural. The facial features of this alii family also seemed to us to deviate significantly from those of the common kanaka, which makes it probable that this whole ruling family is descended from a race of giants, which perhaps even came from another part of the Pacific, and through their strength subjugated the people here. Since the king and the aliis usually intermarry, this race of giants could maintain itself for a long time, although, as we noticed, the women are by far in the majority.

It is interesting to note that Kamehameha I, who brought all the islands of this group under his control, is not a member of this alii family. His facial features show a foreign element and indeed, so it seems to us, an Asian one. We would like to make a connection between this fact and a legend which has been preserved on these islands; namely, that some time earlier, before Cook came upon them, a big ship manned by Spaniards—according to the description—had stopped off at these islands for a long time and that later a Chinese vessel without a mast had also come upon them. The Chinese were supposed to have settled on the eastern side of Maui and their influence upon the inhabitants of that area is supposed to be very striking. In any case it is no longer subject to doubt that the Sandwich Islands were known to the Spanish seafarers long before Cook's voyages, as was convincingly shown by Marchand * and Alexander von Humbolt. **

*Marchand, Etienne, *Voyage autour du monde . . .Tomes 1-4, Paris 1796-1800.* Tome I, p. 416.
**Von Humboldt, Alexander. *New Spain IV.* p. 340.

To this must be added the facts which Captain King recorded in this respect in his account of Cook's third voyage. The Spaniards certainly had the most pressing reasons to conceal all their discoveries in the Pacific, since the damages which they incurred each year from English corsairs and buccaneers was immeasurable. When the time finally came when all secrecy on this point was useless and the Spaniards were forced to open their archives, the glory of the Spanish nation had faded. Now they are being blamed for this concealment, though very likely without reason, for, aside from the English nation, no other nation spent such enormous sums for scientific explorations as the Spaniards. That the fruits of these explorations have so slowly come to light is not the fault of the Spaniards but of those nations which constantly brought war to Spain and thereby also affected the revolt in America.

Kamehameha could well be an offspring of this alii family and those Chinese, whom we mentioned above. In the picture by Kotzebue* we find very striking similarities in his face with the still living Kaiki'oewa, present Governor of Kauai, who stood by Kamehameha as a friend in his conquest of the islands and, as he himself told us, was the Blücher of his army. It could well be that common descent formed the basis of a close bond between these two men, who felt themselves above the ordinary people. Kaiki'oewa is also much smaller than the other members of the royal family. His face has a lighter complexion, his forehead is broader, his nose is flatter, his eyes are smaller and set similarly to those of the Chinese. Kauike'aouli, the present ruler, as well as his sister, the princess Nahi'enaena, whose picturesque beauty is so exaggerated in Lord Byron's *Voyage to the Sandwich Islands,* are also small and are not pure descendants of the royal family. They are the children of the old Queen Keopuolani (died 1823), whom she had by her lover, a common kanaka by the name of Hoapiri [Hoapili]. In earlier times the women of the aliis had the right to possess several men, as well as the men several women. Thus Keopuolani kept her Hoapili when she was married to Kamehameha I after her capture.

*Exploration Voyage in the South Seas. Weimar 1821. Vol. II. p. 15.

The reasons are known why Kamehameha I left the island of Hawaii and made his home on Oahu. Honolulu, the royal residence on Oahu, is now the capital of all the Sandwich Islands. In the year 1831 it had a population of about 7,000 and consisted of 750 to 800 Indian homes, among which stand some 30 very nice houses, usually two-story, which were built according to European tastes. One is considerably surprised when one sees how much nicer and cleaner the homes of the foreign merchants are here than the homes of the merchants in the ports of Peru and northern Chile. The houses of the missionaries stand at the eastern end of the city and are very beautiful. Just at that time they were building a very large stone house the construction of which was of exceptional elegance and durability. The homes of the missionaries compare with those of the Indians as in our country the palaces compare with the ordinary homes of the poorest class of people. Of course the palaces on Oahu look somewhat different than the palaces in Berlin, London, or Petersburg. Even the homes of the royal family are extremely humble compared with the the stately homes of the missionaries, which are most elegantly furnished. The missionary homes are a sharp contrast to the little huts in which Mr. Stewart once lived and which he described in his journal as being so wretched.* Nowadays one finds varnished floors, the finest furniture, and beautiful pianos in the missionaries' homes, and magnificent oil paintings adorn their walls and rooms. Who gave the missionaries, who came to the Sandwich Islands as very poor people, the money for such luxury? Let us not even speak of the capital, which some of these masters are supposed to have accumulated and invested in North America. We think we have reason to suspect that this money was taken from the land and the people for whose enrichment the missionaries had been sent to the Sandwich Islands. Honolulu is at the same time headquarters for the missionaries on the Sandwich Islands and has a church which can comfortably hold 4,000 people.

*Note: *Loc. cit.*, p. 150. Mr Stewart adds: "but that you may see, that ours is no princely establishment, in which we dwell in luxury, and lay up treasures for our children, from the charities of the church" etc.

It is a known fact that the government of the Sandwich Islands has put itself under the protection of Great Britain and is therefore completely secure against all attacks from the outside. Nevertheless, the island government has built fortifications and strongholds and outfitted them with numerous heavy cannons which cost them an enormous sum of money, yet are useless to them. The fort of Honolulu is situated close to the docking place of the harbor. It is surrounded by walls of coral backed by narrow earthen dikes which are so weak, that they could never withstand a cannonball. The embrasures are outfitted with about 24 cannons of unequal caliber. Shortly before our stay in Honolulu they bought a very old, large French cannon, the Marechal de Humieres, from 1680. This cannon weighs 5240 pounds and was brought by a merchantman. It was just ready for emplacement when one morning they found it spiked.

Mr. Adams, of whom we frequently spoke at the beginning of this chapter, lives as Governor of Oahu in the middle of that fort and from there he governs quite as he pleases. He had previously been Governor of Hawaii and when, after Boki's departure, he came to Oahu, he brought along his own soldiers who are presently the plague of Oahu. With the greatest impudence these people allow themselves the crudest caprice, often on the open street. They often take food and other things from the poor, defenceless Indians. For the Indians there is no recourse. They cannot ask for protection against them. In Honolulu they have established a military police which roams about the streets of the city day and night. So one sees the Governor's soldiers prowl around, usually in twos, and one must laugh at them because, except for the malo and a piece of white material over their shoulder,[†] these soldiers are naked. They carry their pouch and musket on their bare body.

After eight o'clock in the evening the kanakas must be off the streets of the city. Those who violate this rule are apprehended by the passing soldiers and, unless they can give them money, are invariably imprisoned in the fort. The whole nation grumbles about these excessive measures, which were introduced

[†]This would be the short cape known as the *kihei*.

through the misguided piousness of the missionaries. But the nation is too good-natured, deprived of all weapons, and so lulled through idle lying about and listening to the palapala [*paipala, Bible*] that it is not possible for it to rise against this until forced to do so by the greatest necessity.

There are people who are determined to support the activities of the missionaries on the Sandwich Islands (we are not speaking here of mission activities generally) and who have praised the inexpensiveness of the food, as well as all other necessities, very highly. Perhaps these gentlemen are familiar only with the high prices on the markets of London, for they do not seem to know that among peoples whose industrial and scientific development stand on a much lower level, the cost of all necessities should be much lower than in that highly cultivated land. The island of Oahu, as well as the entire Sandwich Island group, is a paradise. All the fruits of the tropics grow here in great abundance and all the animals which were introduced have multiplied in surprising numbers. A wise government, one which would not have to go to the North American missionaries for advice, would certainly soon be able to transform this land, whose location is so especially fortunate, into a rich and happy state. But good advisers are rare on the Sandwich Islands and the few men of exceptional worth who have been on these islands for a long time have in recent times been misunderstood and defamed by the missionaries. Even Mr. Stewart was guilty of this when he wrote in his journal about Don Francisco Marin. It is true that Marin acquired great wealth for himself in the Sandwich Islands but he did it in a way which will still benefit the great-grandchildren of the present generation. As to that, Marin intends to die in the Sandwich Islands and leave all his property there to the children who will continue the cultivation of the land. Marin brought to the Sandwich Islands almost all the valuable plants which formed the basis of the wealth of great nations. They had already begun to expand the sugar plantations; a sugar mill was already in operation. But instead of furthering agriculture, the basis of all prosperity, the missionaries' system suppressed it so that the Indians might not be diverted from their worship and their instruction in reading and writing by working in the fields and factories. What a market for sugar and coffee the merchants of the Sandwich Islands could have found in California and Chile, where a pound of raw sugar often demands a price of 12 Prussian Silbergroschen (2 real da plata). Cotton thrives beautifully on the Sandwich Islands and formerly quite a bit of it was cultivated. Had the missionaries taught the cultivation of cotton in the beginning and had they brought looms instead of printing presses, then the people of the Sandwich Islands would be wealthy today and the cotton cloth would provide them with clothing. They now have to buy cotton cloth at great expense from the foreign merchants or use prepared bark as an inadequate substitute.

Money and precious metals, these representative symbols of trade, are lacking on the Sandwich Islands. The national wealth had consisted of the quantities of sandalwood which these islands once possessed but which has almost completely disappeared in a most unwise manner. This wood and now the fresh produce, to which present-day trade is solely confined, have brought a great amount of Spanish silver into the land. But since the trade in sandalwood has almost completely stopped, it is becoming rarer every day. The Chinese, who use this wood in enormous quantities for their incense sticks, have suddenly received huge shipments of it from different areas of the South Sea Islands and from some Indian islands. This depressed the price of this kind of wood tremendously and already in the year 1831 there was no longer any demand for sandalwood from the Sandwich Islands. The small supply which they had still gotten together on Oahu was left. In December 1831 the price of sandalwood in China was as follows:

Sandalwood from Malabar — 11 piasters a picul (133 1/3 pounds)

Sandalwood from the Sandwich Islands — only 1.50 piasters

With that one can consider this trade as finished for the Sandwich Islands. *

*Note: Recently a very detailed article by Mr. Bennet about the various types, the various values, and the occurrence of sandalwood appeared in London's "Magazine of Natural History," Vol. V, p. 255. We refer the reader to it, as we have made no new observations concerning this matter.

The mats, the manufacture and use of which is very extensive on the Sandwich Islands, are exceptionally beautiful—as we
pointed out earlier on page 53. But their price, compared with
that at Manila, is so extraordinarily high as to make them un-
suitable objects of trade.

So today the inhabitants of the Sandwich Islands are limited
solely to the export of their fresh produce, which they furnish
to the ships that come either to take on fresh water or to spend
the winter. The latter are primarily those who hunt sperm whale
off the coast of Japan and who want to escape the severe winter
storms of that area.

One must marvel at the diversity of the fruits and other
foods which one sees on the market in Honolulu but one must
also wonder at their prices since they are offered for sale by a
people which for the most part still goes about unclothed. At
the market we saw the most beautiful melons and watermelons.
Nowhere else have we found better ones. Beyond that we saw
potatoes, sweet potatoes, shallots, onions, ordinary pumpkins,
corn, various types of cabbage—which here on the Sandwich
Islands has no seed—bananas, coconuts, pineapples, the fruits
of the *Eugenia malaccensis,* limes, beans, figs, pomegranates, cu-
cumber (very good), Spanish pepper, taro root, sugar cane—which
is eaten both raw and cooked—etc. Goats, chickens—whose bones
are usually covered with a black periosteum—turkeys, ducks
(Anas brasiliensis), geese and pigeons were also offered for sale.
Here one finds beef, and there young pigs and sheep, then again
eggs and crabs—large and small—and fish and shellfish are offered
everywhere. The fish we saw here were trout and perch, which
the Indians eat raw, dipped in a little sea water. The amount of
fish which is consumed here daily is extraordinarily great and
catching them occupies a large number of people who can always
be seen in their canoes beyond the coral reef and a little farther
out on the open sea. The Indians fish with very long nets which
they float in the water by means of calabashes (the dried shells
of the *Anguria*). When it is time to draw the net together, many
Indians throw themselves into the water and drive the fish together with loud noises, which they make with movements in the

water. At ebb tide, when parts of the coral reef are exposed, one sees many women walk around on it, looking for shellfish and crabs.

At the market place there is an official with a staff in his hand who levies the tax for the royal family during the sale. The people say that a great deal of arbitrariness is involved in this. To close with this matter we will list the prices of some of the provisions which we took on board our ship in the year 1831:

100 watermelons — 4 Spanish piasters (about 1 4/5 Prussian Silbergroschen a piece)

10 pounds of beef — 1 piaster (4½ Prussian Silbergroschen per pound)

15-20 ears of corn — ¼ piaster (in other words more than ½ Silbergroschen per ear)

pineapple — 3 to 4 reals (da plata) a piece

2 arrobas potatoes — 6 piasters

2 arrobas sweet potatoes — 4 piasters

8 bunches of bananas — 1 piaster (when bought individually one often pays 2 reals a bunch)

100 eggs — 2½ piasters (according to which one egg costs more than one Silbergroschen)

46 cucumbers — 2 piasters

1 turkey — 1-1½ piasters

25 chickens and ducks — 6 piasters and 2 reals

30 pounds of grass — 6 reals . . . etc.

The fault lies first of all in the very limited cultivation of the land, which has of late been neglected even more because the natives were otherwise occupied by the missionaries. But most of all there are the high taxes and the great expense of those necessities which were introduced by the foreigners, primarily the need for European articles of clothing. We already mentioned previously the enormous sums which the government spends on the maintenance of a large number of soldiers, who are of no use to the country. The members of the royal family and the ruler himself have become accustomed to the luxuries of the civilized world and they don't want to be second to the missionaries and the foreign merchants in these things. They drink our expensive wines which sell for an enormous price, for example, 1½ piasters for a bottle of an ordinary red wine in the Honolulu Inn. Through the extensive trade in sandalwood, which the Sandwich Islands enjoyed in former times, the high aliis and the King accumulated great sums of money. These are all gone now but the recipients have not wanted to give up their adopted luxuries. We need only refer to the facts which Mr. Beechey* presented so excellently, in order to illustrate the excessive luxury of a royal family which rules over a land of naked inhabitants. They have wasted thousands of piasters on the most useless things.

At the time of our visit, Governor Adams raised over 3,000 piasters in cash and an enormous quantity of food annually for himself and his soldiers. The people have to pay him according to the size of their property. The King of the Sandwich Islands levies a general head tax as soon as he needs money. This consists of one piaster for every Indian, one half piaster for every woman and one quarter piaster for every child. Usually the tax is levied only once a year, but sometimes more often, and whoever cannot pay in cash, pays in foodstuffs. Besides this tax the King receives his subjects on one day during the year and at that time each one must bring him a present. The foreign merchants pay him 20, 30, even 40 to 60 piasters, each according to his means. Besides these very high taxes for such a poor people, they must cultivate the fields of the aliis and the King. For this purpose the King issues the command through a town-crier, that the people from a certain district of the country or the city should assemble the next day and the following days for a specific job. The Indians go out to this with women and children and the part of the city which they leave behind is placed under a taboo so that no one can enter their houses and steal. At the time of our visit to Honolulu a great wall of coral was being built from the city to the fortification on Mount Puowaina. Thousands of people worked on this project and in almost the whole city, even on the market place, the taboo was being called out. We could not even obtain the necessary fresh provisions for the ship.

*Beechey, F. W. *Narrative of a voyage to the Pacific*...London. 1831. Vol. 2:417.

When one visits the great plain of Honolulu and sees the expanses of beautifully cultivated land in the valleys which open up into the plain of Honolulu, as well as the enormous quantity of food plants which are grown in the Pearl River Valley, one might be tempted to believe that there is a great over-abundance of food here; which, however, is by no means the case. The taro plantations take up an enormous amount of space and yield much less food than our potato fields and our grains. Indeed, we would like to attribute the high prices of fresh produce on the market of Honolulu to inefficient land utilization.

The number of ships which visit the Sandwich Islands is increasing every year with the expanding trade in the Pacific. The location of these islands in the northeast trades and on the latitude of Canton is so fortunate, that almost all ships which sail from America—North America as well as South America—to China go by way of the Sandwich Islands and stop off in the Port of Honolulu in order to take on fresh water, some fresh fruits and some livestock. The ships which hunt whale and sperm whale, however, customarily remain in the harbor of Honolulu during the winter months—October through December, when there are severe storms off the coast of Japan—and buy up great quantities of provisions with cash. Also, every ship which puts into the Port of Honolulu must pay a tax of 80 piasters to the King. Those ships which remain in the roads do not need to pay. In this way a considerable amount of cash comes into the Sandwich Islands but it is the only revenue which they enjoy. We believe, however, that it is a great mistake to judge the growing importance of the Sandwich Islands by the increasing tonnage of the ships putting into port there each year. This has probably been done in recent times, but the trade which these ships carry on here is extremely limited, since the Indians lack not only money but also exportable articles. Everything is confined to the sale of fresh provisions.

In the year 1832 the following number of ships came to Honolulu and put into port there: 23 British, 128 American, and 8 ships of other nations, with a total displacement of 41,744 tons. This is approximately as much as the British East India Company once transported in tea alone from China to England.

In regard to the missionaries we would like to mention one more fact which seems to be of importance. As we had already learned back home, several Catholic missionaries, craftsmen, manufacturers and colonists came in the year 1827 from France to the Sandwich Islands in order to teach the Catholic religion and European culture and thereby bring a general state of well-being to these people. The fate of this remarkable expedition recently became known through Mr. Morineau.[*] Kauike'aouli gave these missionaries permission to remain on the Sandwich Islands and to preach the Christian religion according to their principles. These men soon met with great approval, partly because they had more engaging personalities than the North American missionaries and partly because the forms of the Catholic worship service were more appealing to the Indians and its whole nature more fitting to new converts than the service of the Protestant religion. In the Protestant church the common Indians lie on the ground and are terribly bored. Now the North American missionaries have finally carried things so far that the French are forbidden to practice their religion in public. They are treated like prisoners in their homes and their every step is watched. We do not need to point out the errors which the North American missionaries committed in this matter. What have things finally come to after 300 years? The Protestants now deny the Catholics the practice of their religion, a religion which would surely make the Indians, who are still on such a low level of civilization, much happier than our Protestant religion. It would be completely out of place here to discuss the principles according to which religion must be taught to a people who are on as low a level of civilization as the Sandwich Islanders. It is certain, however, that the North American missionaries have taken the wrong way. If only they would become aware of the fact—which Mr. Beechey had already pointed out to them—that the Indians must not in any way neglect the care of their worldly possessions for the purpose of securing a place in heaven after their death.

*Berghaus Annals 1823 p. 1. [Morineau, Auguste de, Notice historique sur les Iles Sandwich 1779-1833, Poitiers; 1834 p.25].

Since some of the North American missionaries are very clever and even know that Prussia is close to France, they tried everything to rid themselves of their dangerous opponents—the Catholic missionaries—by sending them back home on the *Prinzess Louise*. The government of the Sandwich Islands sent a written request to Captain Wendt, even asking for free passage for the missionaries. Captain Wendt rejected this request absolutely, especially since these missionaries declared that they wanted to stay on the Sandwich Islands until they were driven out by force.

This is, then, what we considered important to relate about the Sandwich Islands in general and about the missionary situation in particular. The missionaries, and especially the people who blindly defend the mission establishment, will quickly rise against it and when Mr. Stewart publishes his next journal about his travels as a clergyman—as he is in the habit of doing—he will use every means to perhaps prove some of our statements to be false. This is the manner in which the missionaries have been defending themselves and if these authors once succeed in finding some statement among the accusations to be inexact, they rejoice like children and triumph, as though they had cleared everything. One need only read Mr. Stewart's 58th letter* in which this sort of argumentation approaches the ridiculous, while in fact everyone on Oahu says that Kauike'aouli had wanted to marry his sister in accordance with the ancient customs of the land. We could cite some interesting facts concerning the extent to which the people observe other ancient customs, quite without the missionaries' knowledge. We have not paid any attention to the slanders of the foreign merchants and the doctors, who told us quite unbelievable things in Honolulu, because these have come from all parts of the world. Indeed, there are even people among them who have escaped the hands of criminal justice. There are, however, also good and hard-working men among them. There were very malicious rumors afoot about the illness of King Kauike-'aouli's sister, who was living on Maui in the home of a missionary. No foreigner had seen her for several months.

*A *Visit to the South Seas.* London 1832. Vol. II. p. 190 etc.

As sad as the picture is in many respects which we have had
to paint of the conditions on the Sandwich Islands, we are yet
extremely happy to be able to end this chapter with an indication of the great change which is probably now taking place on
those islands. The political papers have reported the news that
Ka'ahumanu, the old queen and sovereign, passed away in June
of 1832 and that Kauike'aouli, the young King, was crowned as
Kamehameha III and recognized as such by the British and
that he has taken over the government alone. Kauike'aouli has
lifted several of the bans on luxuries imposed by the old Ka'ahumanu and the dances and favorite games of the natives—throwing the javelin, etc.—are permitted again. The Indians who wish
to continue to attend the Christian services are allowed to do so.
Coercion in this respect is no longer tolerated, however, a matter
in which, as in many other things, they say the old queen was
too much influenced by the well-meaning but too uncompro-
mising missionaries.

Hopefully the Sandwich Islands will now quickly move towards the prosperity and importance which one could wish for
them in their fortunate location.

Finally we want to report here the very good thermometer
observations which Mr. Reynold, owner of the Oahu Hotel, recorded at our request, after we had compared his thermometric
with ours and had ordered the necessary precautionary measures
for correct observation. Our short stay in this very interesting
place does not allow us to give a general description of the climate
of this island. We will in time, however, use the occasional observations of other travelers to do a comparative study of the climate
of this island and that of southern China and of Cuba.

BOTANICAL NOTES

1. *Edwardsia chrysophylla* is now identified as *Sophora chrysophylla*, the **māmane**, which is endemic to the Hawaiian Islands and not introduced from Tahiti.

2. *Arum macrorrhizon* is a misidentification of the *Colocasia esculenta* var. *antiquorum*, or **taro**.

3. *Hydrocotyle interrupta* is now identified as *Hydrocotyle verticillata*, the **pohe**.

4. *Jussiaea augustifolia* is now identified as *Ludwiga octivalvis*, the **kāmole**.

5. *Potamogeton* — pondweed.

6. *Arum macrorrhizon* — see note 2.

7. *Caladium esculentum* is a misidentification of the *Colocasia esculenta* var. *antiquorum*, or **taro**.

8. *Arum macrorrhizon* — see note 2.

9. *Convolvulus batatas* is now identified as *Ipomoea batatas*, or the **'uwala**.

10. *Tephrosia piscatoria* and *Tephrosia toxicaria* are now identified as *Tephrosia purpurea*, the **'auhuhu**.

11. *Ipomoea bona nox* is now identified as *Ipomoea alba*, the **koali-pehu**.

12. *Ipomoea cataracta* is probably a misidentification of *Ipomoea congesta*, the **koali-'awa**.

13. This *Zingiber* would be *Zingiber zarumbet*, the **'ōpuhi**.

14. *Curcuma longa* is a misdetermination for *Curcuma domestica*.

15. *Solanum nigrum* — the **pōpolo**.

16. *Musa* — **mai'a**, or banana.

17. *Ipomoea bona nox* — see note 11.

18. *Ipomoea palmata* is now identified as *Ipomoea cairica*, the **koali**.

19. *Jambosa malaccensis* is now identified as *Eugenia malaccensis*, the **'ōhi'a-'ai**, or mountain apple.

20. *Rhynchospora castanea* is now identified as *Rhynchospora larvarum*, the **kuolohia**.

21. *Cyperus auriculatus* is now identified as *Cyperus ferax* var. *auriculatus*, the **pu'uka'a**.

22. *Cyperus owahuensis* is now identified as *Cyperus javanicus*, the **'ahu'awa**.

23. *Panicum pruriens* is now identified as *Digitaria pruriens*, the **kūkaipua'a**.

24. *Aleurites triloba* is now identified as *Aleurites moluccana*, the **kukui**, or candlenut tree.

25. *Dracaena terminalis* is now identified as *Cordyline terminalis*, the **ti**.

26. *Acacia heterophylla* is a misidentification of *Acacia koa*, the **koa**.

27. *Scitamineae* — a plant family which includes gingers and bananas.

28. *Urticaceae* — the nettle family.

29. *Böhmeria albida* is now identified as *Böehmeria grandis*, the **'akoka**.

30. *Aleurites triloba* — see note 24.

31. *Peperomia verticillata* is a misidentification of *Peperomia sandwicensis*.

32. *Cyrtandra ruckiana* is now identified as *Cyrtandra grandiflora*.

33. *Blechnum fontanesianum* is now identified as *Sadleria cyatheoides*, the **'ama'uma'u**.

34. *Aspidium exaltatum* is now identified as *Nephrolepis exaltata*, the **okupukupu**.

35. *Polypodium pellucidum* — the **a'e**.

36. *Hydrocotyle interrupta* is now identified as *Hydrocotyle verticillata*.

37. *Metrosideros polymorpha* — the **'ōhi'a-lehua**.

38. *Dracaena terminalis* — see note 25.

39. *Peperomia verticillata* — see note 31.

40. *Plantago quelcana* is now identified as *Plantago princeps*, var. *queleniana*.

41. *Atriplex oahuensis* is now identified as *Chenopodium oahuense*, the **'āheahea**.

42. *Jambosa malaccensis* — see note 19.

43. *Argemone mexicana* is a misidentification of *Argemone glauca*, the **pua-kala**.

44. *Sida ulmifolia* is a misidentification of *Sida fallax*, the **ʻilima**.

45. *Aleurites triloba* — see note 24.

46. *Convolvulus bona nox* is now identified as *Ipomoea alba*, the **koali-pehu**.

47. *Convolvulus palmatus* is now identified as *Ipomoea cairica*, the **koali**.

48. *Acacia heterophylla* — see note 26.

49. The *Böhmeria* mentioned here would be *Boehmeria grandis*, the **akoko**.

50. The "Olana" mentioned here would be *Touchardia latifolia*, the **olonā**.

51. *Vaccinium cereum* is now identified as *Vaccinium dentatum*, the **ʻōhelo**.

52. *Coffea mariniana* is now identified as *Psychotria mariniana*.

53. *Myonima umbellata* is now identified as *Canthium odoratum*, the **alaheʻe**.

54. *Anoda ovata* is now identified as *Sida fallax*.

55. *Wiegmannia glauca* is now identified as *Hedyotis schlechtendahliana* var. *glauca*.

56. The sweet potatoes mentioned here would be *Ipomoea batatas*, or **ʻuwala**.

57. *Scirpus meyenii* is now identified as *Scirpus validus*, the **ʻakaʻakai**.

58. *Eleocharis palustris* — Meyen was probably referring here to *Eleocharis calva* var. *australis*, or **kohekohe**.

59. *Convolvulus ovalifolius* is now identified as *Jaquemontia sandwicensis*.

60. *Gouania integrifolia* is now identified as *Gouania meyeni*.

61. *Euphorbia cordata* is now identified as *Euphorbia degeneri*.

62. *Cocos nucifera* — the **niu**.

63. *Böhmeria albida* — see note 29.

64. *Cactus ficus indica* is a misidentification of *Opuntia megacantha*, the **pā-nini**.

65. The *Ricinus* mentioned here would be *Ricinus communis*.

66. *Cactus ficus indica* — see note 64.

67. *Arum macrorrhizon* — see note 2.

BOTANICAL INDEX

GENERAL INDEX

For plant references, please see the BOTANICAL INDEX on pages 84–87.